THE
AGELESS
BRAIN

ALSO BY DR DALE E. BREDESEN

The End of Alzheimer's

The End of Alzheimer's Program

The First Survivors of Alzheimer's

THE
AGELESS
BRAIN

How to Sharpen and Protect
Your Mind for a Lifetime

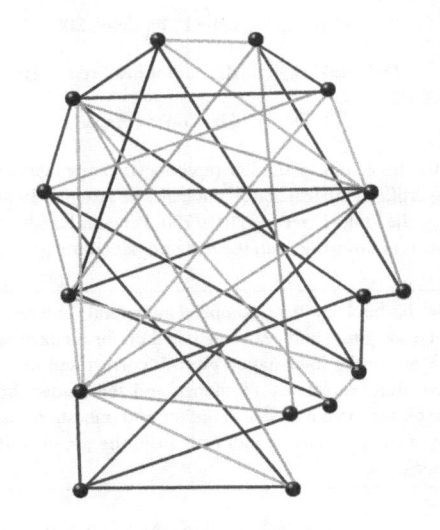

DR DALE E. BREDESEN

Vermilion
LONDON

VERMILION

UK | USA | Canada | Ireland | Australia
India | New Zealand | South Africa

Vermilion is part of the Penguin Random House group of companies whose addresses can be found at global.penguinrandomhouse.com

Penguin Random House UK
One Embassy Gardens, 8 Viaduct Gardens, London SW11 7BW

penguin.co.uk
global.penguinrandomhouse.com

Penguin Random House UK

First published in the United States of America by Flatiron Books in 2025
First published in Great Britain by Vermilion in 2025

3

Copyright © Dale E. Bredesen 2025

The moral right of the author has been asserted.

Designed by Omar Chapa

No part of this book may be used or reproduced in any manner for the purpose of training artificial intelligence technologies or systems. In accordance with Article 4(3) of the DSM Directive 2019/790, Penguin Random House expressly reserves this work from the text and data mining exception.

> The information in this book has been compiled as general guidance on the specific subjects addressed. It is not a substitute and not to be relied on for medical or healthcare advice. So far as the author is aware the information given is correct and up to date as at December 2024. Practice, laws and regulations all change and the reader should obtain up-to-date professional advice on any such issues. The author and publishers disclaim, as far as the law allows, any liability arising directly or indirectly from the use or misuse of the information contained in this book.

Typeset by Jouve (UK), Milton Keynes

Printed and bound in Australia by Griffin Press

The authorised representative in the EEA is Penguin Random House Ireland,
Morrison Chambers, 32 Nassau Street, Dublin D02 YH68

A CIP catalogue record for this book is available from the British Library

Hardback ISBN 9781785045677
Trade Paperback ISBN 9781785042287

Penguin Random House is committed to a sustainable future for our business, our readers and our planet. This book is made from Forest Stewardship Council® certified paper.

This book is dedicated to the fraction of medical researchers and practitioners—physicians, scientists, naturopathic doctors, nurse practitioners, neuropsychologists, health coaches, and others—with the courage, insight, and compassion to study, practice, and advance personalized, twenty-first-century medicine that is focused on patient outcomes rather than outdated practices. On behalf of the many whose lives you have saved, thank you.

CONTENTS

Introduction		*1*
1:	Performance and Protection	9
2:	Adding Insults to Aging	30
3:	What Is Possible at One Hundred and Beyond?	50
4:	Dying of Profit	68
5:	Identify Your Why	85
6:	A Measured Approach	102
7:	Eating for an Ageless Brain	120
8:	The Brainspan Workout	150
9:	Cleanse and Restore	174
10:	The Brain's Flex Factor	199
11:	The Toxic Adventure	220
12:	The Microbial Mind	241
13:	Soonism	260
14:	Prescription for an Ageless Brain	281
Acknowledgments		*317*
Notes		*321*
Index		*361*

THE AGELESS BRAIN

INTRODUCTION

Time is the broker through which we trade our youth and vitality for experience and wisdom. Essentially the deal is energetics for information. But it turns out that the exchange rate—which is our biological aging rate—varies dramatically: some of us will reach one hundred filled with knowledge and capability, whereas others of us will get to forty, look like we're seventy, and already struggle to remember our phone number.

So the key question is: How do we obtain the best exchange rate? In other words, how do we keep our brains youthful, healthy, and functional for the entirety of our lives? Until recently, the answer has been a shoulder shrug—"No idea"—but our armamentarium against brain aging has grown dramatically in just the past few years, and that is what this book is all about.

Starting right here and now, with just a few changes to the way you care for your brain, you can have sharper thoughts, crystal-clear memories, the ability to learn new information easily, and the capacity to take better control of your moods and emotions. What's more, you can retain these powers for one hundred years of life or even longer.

That might seem like an awfully big claim. And you know what? It is! What you will read in this book, after all, is not yet part of mainstream medicine.

I do want to be up-front about that. The information contained in these pages is not what you would hear from most doctors if you were to ask them how to prevent brain aging and disease. What they are likely to tell you is that brain aging is inevitable, and that neurological disease is nothing short of fate if you've inherited the wrong kinds of genes. Alas, *that* is what is widely accepted.

I am among a small but quickly growing number of physicians, however, who no longer think about the human brain in this way, for we know that what is widely accepted is not always what is true.

For example: It was once widely accepted that, as we age, our hearts simply weaken—exhausted from a lifetime of reliably contracting and expanding some 2.5 billion times across our lifetimes, until this organ can simply beat no more. Today, we know that we can do many things to strengthen our hearts and lengthen their usefulness. Some of these decisions—diet and exercise and the like—should start early and continue throughout our lives. Some measures—supplementation and medication and some medical procedures—can come along later and still be effective. And some interventions—bypass surgeries and transplants—are there for when things go very, very wrong. But I would argue there is virtually nobody who still believes nothing can be done to extend the usefulness of a human heart.

It was once widely accepted that, as we age, our cells become more prone to mutation, and these mutations become more prone to spread. Eventually, the logic went, our healthy cells would be outnumbered, and this would be the beginning of an inevitable and often very painful end. Today we know that there are many things we can do to prevent this problem, which is called cancer: to catch it early, to slow its growth, and in many cases to vanquish it. As is the case for our hearts, there are things we should be doing early in our lives, interventions that can happen a little later on, and emergency measures that can be taken when things get very bad. But there is virtually nobody who still believes nothing at all can be done about cancer.

It was once widely accepted that, as we age, our bodies become less efficient at turning food into energy. Eventually, we run out of our stores of insulin, the key hormone for regulating the sugar in our blood, decreasing the elasticity of our blood vessels, impeding the work of our cardiovascular system, hampering the ability of our bodies to heal from wounds, preventing our kidneys from filtering waste, and destroying our nerves' ability to send signals from one part of the body to another. Today we know that there are many things we can do to prevent and reverse diabetes. And as with the heart and our cells, we can take preventative measures early in life, perform procedures as needed at points along the way, and make crisis interventions when catastrophe strikes.

But then there is the matter of our minds. It was once widely accepted that, if you live long enough, you will inevitably face some cognitive decline. Indeed, the Ayurvedic physicians of three thousand years ago considered dementia not as a disease, but as a normal consequence of aging. Once it occurs, it has always been agreed, there is a gradual slope of slowly and ever-worsening memory loss and confusion, with no hope for recovery—a terminal illness. And for some unfortunate souls, the slope is not so gradual. Those with Alzheimer's disease and other dementias would fall into a disorienting hole of memories and dreams mixed with the present in confounding combinations, the world around them no longer recognizable, the future no longer imaginable, their own children no longer identifiable as the most important people in their lives.

But today we know that the loss of these wonderful abilities we have as human beings—to speak, to read, to learn and remember, to calculate, and to reason, among so many others—is not ineluctable, and, in fact, is rapidly becoming both avoidable and reversible. In this golden age, we can identify Alzheimer's disease with a simple blood test, years before there is irreversible brain damage, often before people have any idea they are even developing Alzheimer's. We can identify the reasons—what the contributors are—and most important, we can halt the process and, in most cases, reverse the symptoms. We finally understand the basic principles that underlie cognitive decline, and these have predicted, accurately, how to prevent it and how to treat it successfully.

Unfortunately, most people are still resigned to the notion that cognitive decline is an inevitable fate and that dementia is the destiny of hundreds of millions of people across our world. I've often noted that it is sometimes said that everyone knows a cancer survivor, but no one knows an Alzheimer's survivor.

But let me tell you a secret: I do! I know many of them, in fact.

OK, maybe that's not really a secret. I've tried to let people know. Goodness knows I've tried. I've done everything but shout it from the rooftops, and I would not be opposed to doing that, too, if it would help. Unfortunately, a lot of people are not quite ready to listen yet. This is understandable, for they have been told—correctly so, until recently (2012 to be exact)—that part of growing old is aging, and part of aging is losing our ability to comprehend the world as it is and to remember the world as it was. Thus, part of aging, they believe, is losing who we are.

That, however, is simply not true. We're not nearly so far along in knowing how to treat the neurological complications of aging as we are at treating heart disease, cancer, and diabetes, all of which are still rampant killers in our world. And so, as you can see, when it comes to brain health we are not anywhere near the place we are with other diseases that have been virtually or completely eliminated. Even the most determined optimist in the world doesn't think we're going to end Alzheimer's tomorrow. Anyone who believes so might indeed have been born yesterday, figuratively speaking of course.

But we now know that, like heart disease, cancer, and diabetes, there is so much that can be done—some of it early, some of it a little later, and some of it as a last resort—to prevent, impede, and even reverse neurodegenerative diseases.

This is why people come to us from around the world when they or their family members are suffering with memory and thinking problems, often after they have exhausted all other options, and often when they are feeling desperate and despondent about what is about to happen to them; they have also been told that once they find themselves on that not-so-gradual slope, there is no coming back.

I believed that once, too. But I could not help wondering: Someday, perhaps far in the future, someone will come up with an answer to this problem that has been called "impossible," and develop the first effective treatment for patients with the cognitive decline of aging. What will they or their team think of that we did not? Where will they succeed where we, and everyone else, failed? Will they see this condition differently from the rest of us? Will they innovate to create a novel approach? Whatever they do to succeed, might we figure that out now?

These thoughts made my team and me dig deep, and we spent many years in the laboratory in search of answers. Finally, our research revealed that Alzheimer's (the disease that causes most of the cognitive decline of aging) is quite different from what we had been taught in medical school, so it must be treated quite differently from the standard protocol as well.

That's how we came to develop what we dubbed the ReCODE Protocol (a name first suggested by my colleague Lance Kelly), which is short for **re**versal of **co**gnitive **de**cline. I wrote extensively on this framework for treating neurodegenerative diseases in *The End of Alzheimer's* and its follow-up, *The End of Alzheimer's Program*. In my third book, *The First Survivors of Alzheimer's*, some of the people I have worked with who have used the protocol to slow, stop, and in many cases reverse their slide toward dementia wrote their own remarkable and often poignant stories. But these are the stories of people who were already suffering. And it is my intention that you, dear reader, should never suffer in that way.

Indeed, if you are of the generation termed Gen Z, my belief is that you will never even have cause to worry. There is every reason to believe, and few reasons to doubt, that we now know how to prevent cognitive decline and dementia in virtually everyone. A disease that was the number one worry of people in my generation (also known as baby boomers) is being rendered toothless by scientific research. And in this book, I will describe how we are doing just that.

Despite all this progress, many people still wait far too long to begin treatment—ten or even twenty years after the disease has begun. For those people, there is both good news and bad news. The good news is that we have seen the protocol work wonders: Many memories could return; their ability to recognize loved ones and engage with them often comes back; their speech might return; and their ability to care for themselves may return as well as their capacity to interact meaningfully with the world around them. The bad news is that the later they start treatment, the less likely it is that improvement will occur, and with advanced dementia, the return is partial, not complete—although we do see that it may be sustained for years. The protocol is also more extensive and more difficult for those at late stages.

In contrast, most of those in earlier stages do very well with the protocol, and many return all the way back to normal cognitive function. These are the world's first Alzheimer's survivors, the pioneers who have led the way to a better life for everyone down the line. Therefore, if we could just convince people to see a doctor when their cognition is first affected (or better yet, prior to symptoms), we could truly reduce the global burden of dementia.

This is a book for the many people who are at the beginning of their journey—those who are able to avoid the terrible slope. But we know that, since we can provide a much better life to people at the end of their journey and reverse the decline of people in the middle stages, then if we move upstream, to people who have no symptoms, we should be able to prevent cognitive decline altogether. (Indeed, none of the people on our protocol for prevention has developed dementia.) This book contains everything I believe people can do to avoid hitting the slope at all—to increase their "brainspan," as it were, assuring an ageless, active brain for life.

Consider that possibility for just a moment: a world in which we think clearly, learn, and remember, throughout our lives, without worry. This is my hope for all of us, and I believe it is possible now. It is my greatest wish that this book empower you with information to help you achieve that goal.

1

PERFORMANCE AND PROTECTION

You know what lasts longer than beauty? Being smart.

—ANONYMOUS

Whether you are in your twenties, or your nineties, or anywhere in between, a high-performance brain that is protected from degeneration is your most important possession—especially if it comes with a lifetime guarantee. Until very recently, that has simply not been feasible. But times have changed—dramatically—and in this book I'll show you how to keep your brain youthful for a very long lifetime.

I'll also explain why, if you work proactively to protect your brain, Alzheimer's disease can now be an optional life experience. (Yes, really, and we have the data to prove it.) I'll describe how to enhance the performance of your brain every day. And I'll tell you why the relationship between your brain and your mind is truly, well, *mind-blowing*. So, please strap in, take a deep breath, and enjoy this ride through previously uncharted brain territory. Our destination is Better Living Through Neurobiology. Younger and wiser—that is the goal.

To get there, we must achieve brain performance *and* protection, not simply one or the other.

Performance without protection is easy—cocaine, Adderall, and sugar can all give quick performance bursts at the expense of long-term protection. Conversely, protection without performance is also easy—you could simply have your brain frozen in liquid nitrogen, where you'll have no risk of neurodegeneration but likely won't be pleased with your neural performance. So, the real trick is achieving enhanced performance with a lifetime of protection with no compromises.

But one of the most underappreciated facts about brain performance and protection is that compromise sneaks up on you. As the famous physicist Richard Feynman observed, a compromised brain loses its ability to recognize its own compromise, and indeed Feynman was, at times, a victim of such compromise. In the mid-1980s, while on his way to purchase a new computer, Feynman tripped on a parking bumper and fell into a wall, injuring his head. For weeks afterward his driving skills were atrocious, his lectures were nonsensical, his behavior was disturbing, and yet Feynman didn't recognize that anything was wrong. It was only after his wife demanded that he see a doctor that Feynman learned that he had suffered from a subdural hematoma, which occurs when blood vessels in the space between the skull and the brain are damaged, causing a clot that puts pressure on the brain, which often causes confused or slurred speech, problems walking and driving, and confusion.

"I found it most curious the way I rationalized all of the weaknesses of my brain," Feynman said. "It's kind of a lesson. I don't know what it means exactly, but it's interesting how when you do something foolish, you protect yourself from knowing of your own foolishness."[1]

We do the same thing with brain aging. One of the most common refrains I hear is that "memory loss is just part of normal aging." This belief is both outdated and dangerous, leading many people to delay evaluation and treatment.

That's what nearly happened to my patient Nina, whom I first met at an immersion weekend—an opportunity for people with concerns about brain aging and disease to engage in comprehensive testing aimed at understanding their state of neurological health. Nina had come not because she was deeply concerned about her health, but because she was inquisitive. Indeed, she figured she was being exceedingly proactive.

"There's been a lot of Alzheimer's in my family," she explained to me. "I think I'm probably OK. But there are a few things—just little things. I'm sure it's nothing."

"What do you mean by little?" I asked her.

"You know, the silly stuff," she said. "Misplacing things occasionally. A bit of brain fog. Losing focus. That's just sort of what happens when we get older, right?"

I glanced down at her report. She was only in her forties. "It *can* be," I said, drawing out the verb for emphasis. "But maybe, you know, it doesn't *have* to be. So, let's see if we can figure out what's going on with you."

It has become clear over the years that what is referred to as "normal age-related memory loss" is anything but normal. It's a bit like saying "normal age-related hypertension" or "normal age-related insulin resistance." Are these conditions common? Sure, and tragically so. But that doesn't make them normal. All of these illnesses are reflections of underlying problems that can and should be treated, but in nearly all cases, they could have also been prevented.

The immersion weekend intake process included a version of the Montreal Cognitive Assessment, also known as the MoCA test. The 30-point examination is quick and easy to administer and has been demonstrated in hundreds of peer-reviewed studies to reflect quite accurately a person's slide toward cognitive impairment and dementia, with each question accessing a different part of the brain's capacity to acquire knowledge quickly, process that information, and respond accurately. It's not a perfect tool, but for a ten-minute investment of time, doctors get a pretty good snapshot of the cognitive health of the patient sitting before them. The creator of the test, a Canadian neurologist named Ziad Nasreddine, designed the exam with the intent that a person with a fundamental education—an average high school graduate, for instance—would generally score in the range of 26 to 30 points. A score between 19 and 25 is associated with mild cognitive impairment. And anything below 20—especially when accompanied by problems with daily activities such as dressing, personal hygiene, or the ability to use and care for assistive devices like glasses or contact lenses—is indicative of dementia, a range of conditions characterized by progressive degradation of intellectual functioning, loss of memory, and changes in personality, the most common cause being Alzheimer's disease.

I looked back down at the chart, flipped a page, and found Nina's score.

It was a 23.

The MoCA can be wrong. *Any* test can be wrong. To know with greater certainty what was happening, it was going to be helpful to do some additional cognitive tests, a blood biomarker assessment, and some brain imaging. But Nina's score, combined with the fact that she was concerned enough about her growing forgetfulness and lack of focus that she had decided to come to our immersion group in the first place, gave me reason for grave concern. Her family history of Alzheimer's made it more likely that she was indeed already suffering from neurodegeneration, a process that is commonly thought to lead to irreversible brain damage and cellular death.

When I went home that night I had trouble getting Nina's words out of my mind. Although she had been proactive about coming in for help, she was still telling herself that everything was normal, right up to the point that we met.

"I think I'm probably OK," she had said. "I'm sure it's nothing."

But she wasn't, and it wasn't.

Nina needed help.

A WHISPERING DEMON

Dementia is deceptive.

It's the Grim Reaper dressed in a clown suit. We say the wrong word in a sentence, and everybody gets a good chuckle. We have a "senior moment," and nobody bats an eye. Again and again, I've heard patients describe the ways in which they passed off its first perceptible symptoms as fleeting moments of inattention, distraction, or exhaustion. They forget their keys. They can't remember the name of a coworker. They glance out the window, spot a cat tiptoeing along a fence, and stop to watch it for a moment—then realize that they can't remember what they were doing before. They leave the house and forget where they were headed. They open an email, a simple invitation to a meeting, but they find themselves struggling to understand where the meeting is, when it is, how they'll get there, who else will be there, or what is going to be discussed. We laugh at these foibles—it all seems so normal.

It's true that these things do happen to absolutely healthy people, sometimes. Just as a headache doesn't usually suggest that we are suffering from a brain tumor, difficulties with memory, reasoning, and attention don't always, or even often, mean we're developing dementia. Sometimes cats are just distracting. Sometimes colleagues write unclear emails. Sometimes a clown is just a clown.

I've often observed that this disease whispers into my patients' ears like a demon. "Don't worry," it says. "This is normal. There is nothing wrong. You're overworked. You didn't sleep well last night. You have a lot going on with your family. You're still being productive. You're still a good worker. You're still a good friend. You're still you!"

But all the while, one by one, then score by score, then hundreds by hundreds, the synapses in their brains—the junctions between nerve cells across which electrical or chemical impulses are passed—are losing their structure and function. And it's not until much later that these individuals realize—or more commonly someone around them realizes—that there *really* is a problem.

In this regard, Nina is a great example of why it's so important not to allow ourselves to be deceived by those demons of denial. Though she'd assumed that she was just worrying about "silly stuff," she nonetheless decided to be extremely precautionary because she'd seen what happened to her family members. I wouldn't wish upon anyone the experience of watching loved ones succumb to Alzheimer's, but the awareness that came from that circumstance probably saved Nina's life.

You might have recognized that, just a few pages ago, I emphasized that neurodegeneration is *commonly* thought to lead to irreversible brain damage. I'm uncommon, in this way, as I don't believe that neurodegenerative disease symptoms are irreversible, because I've seen improvements—not only in symptoms but also in cognitive tests, electrophysiological tests, and even MRI brain volumes—many times.

I do think Alzheimers's is tough to endure. I also believe that we don't know nearly as much as we need to know if we are to reverse its symptoms in all cases. But I am confident that we've seen, and published, *many* reversals,[2] and we will see many more in the coming years as individuals begin to recognize that there is reason to hope. Indeed, Julie, one of the patients who has reversed her decline and sustained her improvement for over a decade now, coined the term "false hopelessness" for the bleak view that is being peddled by old-fashioned doctors and foundations that see brain aging and disease as inevitabilities.

In fact, there is plenty of hope, and Nina is just one example. We assessed her, diagnosed her, put her through a rigorous set of medical tests, prescribed an interventionist regimen, and tracked her progress over the next year, by which point her MoCA had improved to a perfect 30. More important, she was no longer reporting any of the experiences that she had once assumed were "just sort of what happens when we get older." She wasn't losing things. She wasn't forgetting obligations. She wasn't struggling to focus on that which was before her.

But Nina is also an example of why we can't wait to notice the symptoms of cognitive decline before we seek help. One of the reasons she stands out in my mind, after all, is that she was unusually proactive.

Performance and Protection 15

Yet by the time she got to our immersion program, the evidence suggests, she was already sliding toward dementia.

But dementia is simply the fourth and final stage of a process that is ongoing for about two decades, so there is ample time to intervene.

The first stage is presymptomatic, a point at which our biochemistry of degeneration has begun to set in but our lives have not yet been impacted. We've been able to identify this stage for many years using tests such as a PET scan or spinal fluid analysis, but there is also now a sensitive version of the blood test for p-tau 217, which we'll discuss in more detail in chapter 6, which can also provide a warning long before symptoms like memory loss set in.

The second stage is called subjective cognitive impairment (SCI), and in this stage, you know that your cognition is not quite right, but you are still able to score in the normal range on cognitive tests. This is what a lot of people think of as "normal" cognitive aging, which is why many people simply expect it to begin at some point in their forties or fifties, but recent research has shown that those with the most common genetic risk for Alzheimer's display memory differences as early as their late teens. Yes, by the time some people are graduating high school, they're already experiencing some measurable cognitive impairment![3] This is one of many reasons why optimizing and protecting cognition is crucial for everyone. For most people, SCI lasts about ten years—and is readily reversible.

The third stage is mild cognitive impairment (MCI), which means that you are no longer testing in the normal range on cognitive tests (which is what Nina's MoCA showed), but you can still care for yourself, and perform your activities of daily living (ADLs), such as personal care, driving, using your cell phone or other devices, and balancing your finances. It's a shame this was dubbed "mild" cognitive impairment because it's at a relatively late stage of the process—telling someone not to worry because they have only mild cognitive impairment is a bit like telling someone not to worry because they have only mildly metastatic cancer. As one patient told me, "There is nothing mild about what was happening to me." About 5 to 10 percent of patients with MCI will convert to dementia each year.

The fourth and final stage is dementia, which means that your ADLs are no longer intact—you may have trouble with driving, calculating a tip, or dressing. Unfortunately, patients with dementia ultimately develop problems with every task of daily life and eventually become unable to care for themselves. The one bit of good news here is that we have seen repeatedly that, even at this late stage, patients can be stabilized and often improved somewhat, for years. But if everyone would get serious about active prevention, or earliest treatment (during SCI), dementia would be a very rare condition. This is not a fantasy—it is available *today*.

Therein lies the rub. Most people are not nearly so proactive as Nina. By the time symptoms do begin—and especially by the time people stop ignoring the demon whispering into their ear—they are often much, much worse off than Nina was.

REDEFINING HEALTHCARE

What this all means is that we should not wait for cognitive decline to become symptomatic before we do something about it. Indeed, I have come to believe that in every aspect of medical care, symptomatic disease should be regarded as a great failure. And yes, I recognize that this is a radical reinterpretation of the purpose and function of healthcare, which has long been focused on treating people who are sick rather than preventing people from becoming ill to begin with, but I'm not the only person who believes this.

Way back in the early 1970s, when I was a student at the California Institute of Technology in Pasadena, the professor who was assigned to liaise with those of us who were considering going to medical school was an MD-PhD named Leroy Hood. Today, Lee is an absolute legend of the scientific world as one of the founders of the Human Genome Project and the developer of the technology that permitted the sequencing of the first human genome. Even then, though, the young professor was a larger-than-life figure on our campus, and I remember being very impressed by his wit and wisdom. His most important advice: Go to a school driven not just by the goal of creating great physicians but also doing great science. "You might eventually focus on research or you might eventually focus on patient care," I remember him saying. "But these are parts of a whole. It is the combination of science and medicine, not either alone, that has changed the trajectories of human health in the past and will continue to do so in the future."

My respect for Lee has only grown over the years, particularly as he has somehow managed to predict and help usher in many of the most important developments in health science. The fusion of engineering and biology. The importance of the genome. The integration of cross-disciplinary research into biology. The growing demand for personalized medicine. Lee was at the forefront of all those shifts. So, when Lee says, "Hey, pay attention, this is about to happen," it would certainly behoove us all to sit up and listen. And starting in the mid-2010s, Lee became increasingly vocal about another coming revolution, which he called "scientific wellness," that starts with the radical premise that if we can identify wellness-to-disease transitions long before people become symptomatic, we can bring an end to almost all chronic disease in the world.[4]

And yes, Lee believes—and I wholeheartedly agree—that Alzheimer's and other dementias, thus far so seemingly impossible to treat, let alone cure, will be among the conditions that we will be able to eliminate in the next decade or two.

But what is even more important, perhaps, is that by systematically attacking the causes of these conditions even further upstream—as close to that wellness-to-disease point of inflection as possible—we would not only be bringing an end to the neurodegenerative diseases but also eradicating all other common forms of cognitive decline, including that which might not ever materialize as a dementia but which we accept as "just sort of what happens when we get older."

That's what this book is about. We can stop these diseases, and we shall. But we're going to do it by ridding the world of the horrendous notion that our minds have to falter *at all* as we get older.

So, yes, we hope that all of us together shall end Alzheimer's—by employing a new, sensitive blood test; convincing people to come in for prevention or earliest treatment; and utilizing the precision-medicine protocol we developed, we should progress a long way toward accomplishing that goal. And as long as we're doing that, we might as well end all other dementias and the other most common neurodegenerative diseases, too. Let's say a very hearty goodbye to Parkinson's disease. Let's eradicate prion disease. Let's never again have to say Lou Gehrig's name in association with anything other than his history-making seventeen seasons and six World Series championships with the New York Yankees, because we have eliminated amyotrophic lateral sclerosis (ALS, also called motor neuron disease). Let's destroy Huntington's disease. Let's never again worry about spinal muscular atrophy. Let's stamp out spinocerebellar ataxia. If we do all that, it could help hundreds of millions of people.

But even if we were able to end all those diseases at the point of diagnosable symptoms, we'd still not have nearly the effect on the world that we would have if we prevented all forms of cognitive decline for *everyone*.

I understand that this way of speaking about human health may be confusing for those who have come to believe, as Nina did, that confusion, memory problems, and lack of focus are *inevitable* parts of getting older—not really like having a disease at all, in the classical sense, but rather like going gray, getting wrinkles, or complaining about the music kids listen to these days.[5] It's probably even *more* bewildering for those who believe that cognitive decline is simply "baked in" by virtue of the genes we have inherited from our parents. These are common beliefs and, as such, I understand that it might sound strange when I say the fight against cognitive decline is one that everyone should be engaged in, preferably starting with habits we form as children and followed by very purposeful actions in early adulthood, regardless of what genes we carry.

But if we all did that, we could end those diseases for hundreds of millions of people *and* also have brains that operate at full capacity—performance and protection—for as long as we each exist on this planet, for all the other billions upon billions of people in the world.

A TERRIFYING DEVELOPMENT

I do realize how audacious this all sounds. But I also believe it to be true. And it couldn't be more important. Because humanity has been losing this fight.

It's not just that more people are getting old, and so more people are suffering from decline and, ultimately, dementia. It's that—for reasons that aren't yet broadly agreed upon in the scientific community—people are suffering from these conditions earlier and earlier in their lives.

It's been quite a few decades since I followed Lee Hood's advice and headed off to the Duke University School of Medicine. A lot has certainly changed since then when it comes to the ways in which we understand human health and well-being. But nothing strikes me, nor scares me, more powerfully than the fact that back when I was in training we virtually *never* saw people in their thirties, forties, or fifties with Alzheimer's disease, and now we do—quite frequently, in fact.

Early-onset dementias used to be fodder for fascinating single-patient case studies in medical journals. These were the rare exceptions that challenged, but ultimately affirmed, the rule: In the vast majority of cases, the symptoms of Alzheimer's, even the early signs, wouldn't appear until much later in life. These days, however, I no longer feel surprised to meet someone who has been diagnosed with early-onset Alzheimer's at a time in life that is usually reserved for growing careers and building families.

This is not just my anecdotal experience. Researchers from the Blue Cross Blue Shield Association, the umbrella organization for dozens of locally operated companies that provide health insurance coverage for about a third of Americans, were also taken aback by the seeming growth of early-onset dementia and Alzheimer's. They resolved to track the phenomenon more closely. In 2020 the organization released the startling results of that effort. Yes, the report's authors affirmed, the vast majority of cases are found in Americans over the age of 65. But between the years of 2013 and 2017, the researchers found, there had been a *143 percent increase* in diagnoses among those between the ages of 55 and 64.

And a 311 percent increase in diagnoses for those between 45 and 54.

And a 373 percent increase in diagnoses for those between 30 and 44![6]

Among that latter group, the researchers estimated that nearly twenty thousand Americans in their thirties and early forties had been diagnosed with early-onset Alzheimer's in a single year.

This was a review of commercially insured Americans. What that means is that it likely underreported the prevalence, since income and employment—the two main determinants for health insurance—are associated with disparities in rates of Alzheimer's disease.[7]

If this was *just* one report, even I'd be skeptical. But it's not. It's part of a growing body of literature suggesting that early dementia is either a quickly growing problem or a challenge that we've been dealing with for a long time without really realizing it.[8] I suspect it's a little of both.

We are now clearly better at catching neurodegenerative disease than we were decades ago. Today, simple cognitive tests are freely available online, so many people are beginning the diagnostic process long before they see a physician; we've essentially lowered the barriers to initial screening. Meanwhile, blood biomarkers have joined brain scans as a minimally invasive way to know whether an attribute such as increasing forgetfulness might, in fact, be a symptom of disease.[9] These changes are likely to be at least a part of the skyrocketing diagnosis increases.

Performance and Protection

But that's not the whole story. Some researchers believe that excessive television, computer, and mobile-device screen exposure during critical periods of brain development can lead to cognitive impairments in early adulthood that result in substantially increased rates of early-onset dementia later in life.[10] Other health effects related to excessive screen time have been well documented in individuals born after 1980—and that's the very age range in which we are seeing such profound increases in the past few years. Obesity has also been linked to cognitive deficits, brain atrophy, and impairment of synaptic activity,[11] and the rate at which this condition is present in younger people has been skyrocketing in the past few generations as well. It is on top of this research that I offer my experience as a physician who has evaluated thousands of patients. To me, there is simply no question: People are becoming symptomatic earlier than ever.

Yet, for the most part, doctors treat young patients the same way they treat very old ones. They say they are very sorry, but not much can be done. Neurodegenerative disease, they often say, is little more than being dealt "a bad hand." And while there are some medications that have shown a modest degree of success with some people in some instances, no honest doctor would ever profess optimism that these drugs will work on any given patient, nor that they will do much more than transiently enhance cognition or slightly slow down the progress of the disease. So, at best, this is a palliative approach, aimed at mitigating suffering for as long as possible. At worst, though, it is just hospice in slow-motion, starting with the assumption that the only conceivable outcome is a continual and escalating loss of neurological function, replete with mood and personality changes, confusion, suspicion, depression, fearfulness, and outright terror.

For years, I've said that messages like these—directed at people in their sixties, seventies, and eighties—are a form of cruel but all-too-usual punishment for the mere sin of getting sick. But now, after the improvements we have seen and published (and others have begun to publish as well), when doctors tell people in their thirties and forties—or any age, for that matter—that there is no hope, it's not just cruel. It's malpractice.

And, at least in my mind, it's not a far step from "you have been diagnosed with a neurodegenerative disease and there is no hope" to "you will grow old and begin forgetting things, have trouble focusing, and struggle to grasp new concepts." One of these messages just happens to be directed at a segment of the population, while the other is what we tell everyone.

It's truly unconscionable. This horrific nonsense has gone on long enough, and we should not put up with it any longer.

Aging doesn't have to come along with *any* cognitive decline. Certainly not in our fifties and sixties, as is quite common, but not in our seventies or eighties, as is fully expected and accepted, and not even in our nineties or when we turn the corner on a hundred years of life. We don't have to experience brain aging and neurodegenerative diseases at the current rate. We can have protection and performance for life.

To understand how this can be, we must get to the very crux of what robs us of optimal function as we age. And that has everything to do with evolution, so we need to go back in time—quite a way back, in fact, to the foundations of life on this planet.

EVOLVED TO THE EDGE

There are quite a few competing theories about how life on Earth began. Some scientists believe it all started near volcanically active hydrothermal vents deep in the ocean. Others think it happened because of the heat, acidity, and wet-and-dry cycles of hot springs on land. A few have suggested that it came by way of asteroids that crashed into our planet, sowing biotics and prebiotic compounds across the primordial globe, a theory known as the panspermia hypothesis. It does not matter which of these theories you believe to be most likely, or that you believe any of them, because pretty much every scientist agrees upon what happened *next*.

Competition happened. For food. For battle supremacy. For procreation. Organisms fought with other organisms for resources. Species went to battle with their kin for survival. Those who won continued to evolve, and their descendants are with us today. This is how you get a brilliant human from a bacterium—it just takes a while! Those who lost have disappeared, drowning in the deep end of the gene pool. So, you can see why even a tiny advantage is readily selected by evolution: At the right place and time, very small shifts in physiology made a very big difference for what lived and what died. But the energy that could be dedicated to any change was always finite. Ergo, there were always trade-offs—and at each step of evolution, immediate performance was typically selected over long-term protection. When you live fast and reproduce young, the "selfish gene" lives on.[12]

This selection of performance over protection is a form of *antagonistic pleiotropy*, a concept first introduced in 1957 by the evolutionary biologist George C. Williams, who suggested that genes related to early-life fitness are selected at the cost of later-life decline.[13] This concept has been supported in studies of organisms at every size and scale, such that it is now the most-accepted theory of the evolution of aging,[14] and it can be applied to brain aging as well—particularly for a species that became such a dominant force on this planet because of its penchant for perception and planning. It means if we are going to outthink or outmaneuver our competition, then we are going to be susceptible to brain aging and diseases like Alzheimer's, Parkinson's, and ALS.

In *The Character of Physical Law*, the famed physicist Richard Feynman observed, "Nature uses only the longest threads to weave her patterns, so each small piece of her fabric reveals the organization of the entire tapestry."[15] Feynman was referring to the universality of gravitational law, where a tiny experiment in a laboratory is governed by the same rules as the entire solar system, which in turn follows the same physical principles as the trillions of galaxies in the observable universe. But this theme can be equally applied to biology, where you see the selection process driving aging and neurodegeneration, but you can also see it in tiny cellular organelles like the mitochondria, the "batteries" of our cells. The world's leading mitochondrial physiologist, Prof. David G. Nicholls, once told me that early studies of mitochondria assumed they were built like trucks—durable and dependable. Instead, these organelles turned out to be much more like Formula One racing cars—hyper-finely tuned but very susceptible to damage and dysfunction. Nobody expects a high-performance racing vehicle to last for 500,000 miles!

What is true for our mitochondria is true for our brains as well. Our neural subnetworks are true marvels of genetic engineering. When you stomp on the accelerator in your car, think about the striking power amplification you get—from the modest power you used to press your foot down to the roaring horsepower you received from your car's engine. But you actually get an even bigger enhancement—over ten thousandfold, in the blink of an eye—when you amplify the power required for a single thought (which is about a millionth of a calorie per second, a small fraction of the total brain usage, which is a few thousandths of a calorie per second) to produce the maximum power you can generate using the rest of your body, whether the purpose is fighting for survival or lifting weights at the gym (which is about a half a calorie per second).

To achieve this dramatic, nearly instantaneous amplification, evolution has selected for a system that is literally explosive. Imagine putting jet fuel in your car and driving it two hundred miles per hour, knowing that at any time it could blow up! That's what's happening at a submicroscopic level in our bodies. Your motor neuron system uses an "excitotoxic" neurotransmitter, glutamate, which means that it not only triggers this wonderful amplification from thought to strength but also kills the same neurons it stimulates if it is not removed promptly! Therefore, you are at risk of having this power "network fail"—developing ALS—if you are slow at removing the glutamate or if you just trigger it, and trigger it, and trigger it some more over years, burning out the network. It may come as little surprise, then, that accomplished athletes like Lou Gehrig (who set the record for playing in the most consecutive games), whose name is now synonymous with ALS, are at increased risk, as are those who happen to have mutations in their glutamate transporters and are thus slow to dampen the excitotoxic effect.

Finely tuned systems are subject to dysregulation. And the metastable system that adds and removes glutamate from our neurons is no exception. We've learned that this system can become imbalanced as a result of exposure to lead, herbicides such as glyphosate, pathogens that cause Lyme disease, and a bacterial toxin that mimics glutamate, called beta-methylamino-L-alanine. These are not "causes" of brain aging and disease, per se, but rather "insults" that are like bumps on a track upon which a racecar is speeding along at hundreds of miles per hour.

This same theme recurs again and again. More than 200 million people in the world suffer from macular degeneration. In the United States, one of every ten people over the age of fifty develops this condition. This is yet another neural system operating at the limit and prone to failure with age. The macula, the key part of our visual system for color vision and fine details, demands the highest metabolic rate in the body—anytime light falls on your macula, the photoreceptor cells are activated incessantly. Thus, anything that compromises the supply (such as cigarette smoking, vascular disease, or living at high altitude) or increases the demand (such as prolonged light exposure, blue light, or inflammation) increases the risk for macular degeneration.

All of our supremely tuned neural subnetworks—honed through eons of evolutionary performance selection—are at high risk as we age. That includes the network that mediates our neuroplasticity, which regresses in Alzheimer's disease. It also includes the network that mediates our motor modulation, which degenerates in Parkinson's disease and Parkinson's-related diseases such as progressive supranuclear palsy. It includes the network that mediates our motor power, which degenerates in ALS. It includes the network that mediates our fine central vision—which declines in macular degeneration. Each of these networks has its own unique profile of supply and demand, and its own set of insults that are most likely to dysregulate the system.

But these same Achilles' heels are also our entry points for effective prevention and treatment for all.

SUPPLY AND DEMAND

Conceptually, when it comes to battling neurodegenerative diseases, our job is pretty straightforward: Identify the needed supplies and ongoing demands for each disease (including age-related changes) and then address those needs with a personalized, precision-medicine protocol to ensure the demands are once again met by the supplies.

Practically speaking, there is still some heavy lifting to be done to determine the many different insults at play and the fractional contribution of each. But with patient data and help from artificial intelligence (AI), this is all very feasible. And, in the meantime, we can help many in need with what is already known about these various networks.

For example, the Achilles' heel of the motor modulation network that degenerates in Parkinson's disease has turned out to be a specific set of proteins in the mitochondria—respiratory complex I—whose job is to begin the fascinating conversion of food to supercharge your cellular batteries. Anything that inhibits that process can lead to Parkinson's, and it has turned out that there are several common offenders, including organic toxins such as a degreaser used in electronic manufacture and dry cleaning called trichloroethylene (TCE), the herbicide paraquat, and possibly another common herbicide called glyphosate,[16] among others. We obviously shouldn't wait to develop Parkinson's to worry about these chemicals. It behooves all of us to test for these toxins as early as possible so that we can avoid that and other diseases altogether.

With Alzheimer's, the subnetwork at risk evolved for neuroplasticity—the biochemical function of our brains that allows us to learn new information, using it to change behaviors, and outsmart our food and foes. This network is the storage site for a prodigious mountain of memories. Your brain can store 2.5 petabytes of data (that is 2.5 million gigabytes), which is as much as a few thousand home computers. It is essentially a supercomputer that runs on the amount of energy it would take to light a small electric bulb! But the cost for this high performance is long-term degradation, and so commonly the system that supports the generation and maintenance of these memories begins to fail as we age, most commonly from Alzheimer's disease.

Anything that reduces the supply or increases the demand on this network will increase your risk for degeneration. We'll delve more deeply into each of these topics later in this book, but it's worthwhile to see how supply and demand work together to drive the delicate balance of brains that were designed for immediate high performance at the expense of long-term protection.

Number one on the supply list for brain health is energetics. Blood flow, oxygenation, mitochondrial function (the "batteries" of your cells), and fuel (glucose or ketones) are the key players in determining your risk for Alzheimer's. Thus, it is predictable that reduced blood flow (as occurs with atrial fibrillation, for example), reduced oxygenation (common with sleep apnea), reduced mitochondrial function (as occurs with mercury exposure and other toxins), and reduced glucose utilization (as occurs with type 2 diabetes and insulin resistance) are all risk factors for Alzheimer's. Conversely, addressing these various deficiencies improves cognition predictably, as an international team of researchers demonstrated in a randomized, double-blinded, placebo-controlled human trial of metabolic activators that included L-serine, nicotinamide riboside, N-acetyl-L-cysteine, and L-carnitine tartrate in 2023.[17]

Number two is trophic, which refers to factors that help cells survive and regenerate. Trophic factors come in three types: nutrients (like vitamin D); hormones (like estradiol); and neurotrophins (like brain-derived neurotrophic factor, or BDNF). Predictably, reductions in any of these are associated with Alzheimer's disease, and supplementation is associated with cognitive improvement.

The third factor on the major supply list is neurotransmitters. The most important neurotransmitter for memory is acetylcholine. Again, it is predictable that low intake of choline (vitamin B_4, a building block of acetylcholine, is found in eggs, liver, fish, and cruciferous vegetables, among other sources) is associated with Alzheimer's.[18] It is concerning, then, that most of us have a suboptimal intake of choline.

But what factors drive up the demand placed on the neuroplasticity subnetwork? Number one is inflammation, and it is noteworthy that amyloid, which for more than three decades has been vilified as "the cause" of Alzheimer's,[19] is actually a component of the innate immune system (the older, less specific part of the immune system). Amyloid is an antimicrobial peptide. So, when your brain makes the amyloid beta that characterizes Alzheimer's, it is not trying to *give* you Alzheimer's; it is trying to surround, sequester, and destroy microbes that are dangerous to your brain. Therefore, you can think of Alzheimer's as a network insufficiency driven by inflammation of the brain—caused by an infection or autoimmune response. Making matters worse, this immune response demands energy, so again the balance is tipped toward degeneration. You can thus predict that anything that increases inflammation—from poor oral hygiene to a leaky gut to metabolic syndrome to recurrent herpetic cold sores—will increase your risk of cognitive decline. Just as predictable, identifying and treating these pathogens, along with lowering the inflammation, is beneficial.

The number two demand on the neuroplasticity subnetwork is toxins (which actually check multiple boxes, since they can reduce energetics, cause inflammation, reduce trophic support, impact neurotransmitters, and increase stress), and there are three types: inorganics (like air pollution and mercury), organics (like anesthetic agents and glyphosate), and biotoxins (such as those produced by molds). Hence, detoxification is important for prevention and reversal of cognitive decline. This has proven to be the most difficult part of evaluation and treatment, because there are many toxins, detoxification may take years, and the current standard of care for Alzheimer's-related dementia completely ignores biotoxins as common contributors.

The third and final entry on the major demand list is stress. Common forms and causes of stress like anxiety, depression, and insomnia are all related indicators that the nervous system is being overdriven and not coordinating functionally. We have seen repeatedly that people who do well with treatment take a big step backward when they are stressed, and this is especially true for those in whom toxins are dominant contributors. Red-eye flights, difficult relationships, surgical procedures, accidents—all these often exacerbate cognitive decline. Conversely, meditation, yoga, improved sleep, and other stress-reducing approaches represent an important part of an optimal treatment protocol.

PREVENTING DISEASE BY GETTING WAY AHEAD OF IT

Overall, this new insight into neurodegenerative diseases as network insufficiencies, born of evolutionary selection for performance over durability (the "antagonistic pleiotropy" that George Williams suggested way back in the middle of the last century) and nurtured by the insults of modern living, tells us a lot. It tells us how to gauge risk. It tells us how to evaluate people with symptoms. It tells us why there are such disparate risk factors. It tells us how to reverse disease trajectories.

But perhaps most important, it tells us how to prevent symptomatic disease, through a regimen aimed at preventing both brain aging *and* cumulative insults, way ahead of the wellness-to-disease point of inflection. This completely circumvents the current approach of waiting until the later, harder-to-treat stages (which typically occurs simply because people can't believe that something can be done).

Preventing brain-aging symptoms might seem difficult, even impossible, but the truth is that, with minor modifications, it's a remarkably similar approach to the one that renders us bulletproof to Alzheimer's and other neurodegenerative diseases—the chief distinction being that a diagnosis of Alzheimer's requires some inflammation, since the amyloid itself is part of the inflammatory response, whereas non-Alzheimer's aging-related brain changes focus more on energetic insufficiency. Thus, avoiding Alzheimer's and optimizing brain aging are nearly identical concepts.

If we prevent aging and mitigate the cognitive insults we face throughout life, we won't have to worry about diseases like Alzheimer's. We can have a lifetime of performance—a *brainspan* that lasts for a hundred years or more.

2

ADDING INSULTS TO AGING

> Dismiss that which insults your soul.
>
> —WALT WHITMAN

Aging is inevitable.

At least, that's what we've been told for a very long time. And, with the greatest of respect to scientific colleagues who have proposed that it might not have to be this way, there is still no *conclusive* evidence that we will be living in a post-aging world any time soon.[1] But that doesn't mean we have to experience aging in the way we've come to expect—especially not when it comes to brain aging.

Aging is a biological process that is often correlated with the chronological number of years we spend on this planet. It's not, however, *caused* by those years. We all know people who look and act older than they are. And these days we're coming to meet more people who look and act much younger than they are. So, intuitively, we know that age and aging are not synonymous. Thus, as we will use the term for the rest of this book, aging is not something that happens at a fixed rate; instead, it is the mutable deterioration of physiological functions. Particularly, we will mainly speak of functions that relate to cognition—although, as we will see, it is not helpful to separate our bodies from our minds, as aging in any one place in our bodies will soon affect cognition.

It was long thought that the earliest forms of life did not age. Given, however, that it has been shown that even bacteria experience a condition that looks very much like aging at a molecular level, it is likely that aging originated in some of the world's earliest life-forms due to antagonistic pleiotropy—that ancient trade-off to gain immediate performance at the expense of durability and protection.[2]

Aging is a key driver of the disease-causing imbalances I mentioned in chapter 1. A body that has experienced higher rates of aging, after all, is less adept at processing insulin, less capable of shutting out or expelling toxins, less able to fight off pathogens, more subject to genetic mutations that might not be conducive to survival in the modern world, and less able to adequately produce the neurotrophic factors that regulate every system, including those that control our response to stress, such as cortisol. Thus, aging can be thought of as an "integrative insult," or one that affects all the body's systems.

But it's important to note that cognitive decline, and resultant neurodegeneration, can also happen irrespective of aging—indeed, decline can come simply as a result of the insults (such as infections and toxins) to which we are exposed. Thus, while fighting aging is a very good start for preventing cognitive decline, if we want to live for a hundred years with the ability to think and reason, to remember and problem-solve, to adapt to new ideas and environments, and to make decisions quickly and logically, we cannot only address biological aging. We're also going to have to account for the accumulated insults we suffer along the way.

But here's the thing: Everything we do to protect ourselves against these insults appears to have an effect on aging as well—that is, slowing down integrative insults to the point that they're no longer such a potent risk factor for cognitive decline. Instead, our brains are well equipped to deal with such insults at eighty, ninety, one hundred years old, and beyond.

THE SWEETEST POISON

One of the most common questions I am asked is, "What is the single most important thing to do to avoid brain aging and cognitive decline?"

I always explain that it's not about *one* thing. As we've discussed, there are six major groups—three needs of supply and three burdens of demand—that must be addressed to keep our metastable brains protected while enjoying full performance, and many members in each group, so that each of us has a different set of contributors of imbalance. But there is one thing that impacts all six groups, so if you *had* to choose only one thing that would have the greatest impact—one big bogeyman to defeat—it would be sugar. (We will delve into diet and nutrition in further detail in chapter 7.) Sugar is the biochemical equivalent of a million years of evolution compressed into 30 minutes: Just as our ancestors' DNA was selected over the eons for performance over durability, so we select sugar for a brief burst of energy, at the expense of lifespan and brainspan.

Alas, quitting sugar isn't easy. We're evolutionarily primed to love it because it offers an incredibly rapid burst of energy—the sort of verve our ancestors needed to compete for survival in a world in which short-term performance is evolutionarily prioritized over long-term protection. This wouldn't have been possible if we had not evolved to be such rapid producers of insulin, which enhances the capacity of skeletal muscle to produce mitochondrial adenosine triphosphate (ATP), which is the source of most energy in our cells. In the modern world, though, we've learned to process sugar efficiently, pack it into all manner of edible things, and even sprinkle extra onto and into food and drinks we somehow don't believe are naturally "sweet enough."

We can easily see sugar's impact on aging on the largest organ of the human body: our skin, which often reflects what's going on in our other systems. Glucose and fructose create chemical bonds with the fibers of collagen and elastin, the main structural proteins in our skin, rendering the fibers less capable of repair and making the neighboring proteins brittle. This process, which results in advanced glycation end products (AGEs), is happening throughout the body, including in the brain, where AGE receptors play a role in amyloid's inflammation and negative impact on memory. But we can see it most obviously in the skin, in part because it is further aggravated by ultraviolet light.[3]

If sugar only accelerated aging, it would be bad enough, but it also independently accelerates disease.

Most people, if they were asked to think about the chronic consequences of consuming too much sugar, would likely think about diabetes, which they might associate with a variety of bodily symptoms, like blurry vision, numb limbs, exhaustion, dry skin, slow-healing sores, and rampant infections. They're not wrong. Right now, diabetes affects nearly 500 million people across the globe and is expected to rise to about 725 million by 2050.[4] But when I think about sugar, my thoughts immediately turn to the absolute havoc this *poison* (yes, I'm firmly on the side of people like pediatric hormone disorder specialist Robert Lustig, who has been using that word to describe sugar for decades) has on our brains.[5]

Insulin resistance, a reduced biologic response that occurs when cells are repeatedly stimulated by insulin and down-regulate their ability to respond to it, has a profound influence on the ways in which brain cells transform sources of fuel into work. When insulin signaling is compromised, cell communication and survival is compromised. Insulin deficiencies are also known to limit the healthy functioning of the tens of thousands of different peptides and proteins in the human body, including amyloid beta and tau. The presence of amyloid beta and tau deposits are two well-known hallmarks of Alzheimer's disease.[6]

Indeed, there are multiple direct links between sugar and Alzheimer's.

First, when glucose spikes your insulin, your body must then destroy the insulin to avoid hypoglycemia (low blood sugar, which damages the brain), and it does this with insulin-degrading enzyme (IDE), which in a healthy functioning brain also degrades amyloid. With high levels of insulin, amyloid builds up, because the IDE is too busy degrading insulin!

Second, the signature of Alzheimer's on the classic PET scan (the fluorodeoxyglucose PET) is the result of insulin resistance: Reduced glucose utilization occurs in the temporal and parietal lobes, so when the radioactive tracer is scanned, the pattern looks like an L on each side of your brain, running along your temples and up from your ears.

Third, type 2 diabetes, prediabetes, pre-prediabetes (insulin resistance), and metabolic syndrome—all conditions of reduced insulin signaling—*all* increase your risk for Alzheimer's. In fact, the Alzheimer's amyloid blocks insulin receptor signaling, so amyloid production is not only a protective response to microbes but also a regulatory response to high insulin levels.

Fourth, insulin is one of the trophic factors for neurons, so loss of insulin signaling due to insulin resistance pulls the rug out from under our neuronal support.

Fifth, glucose, in an event called "non-enzymatic glycation," attaches to many proteins (including hemoglobin, which is why A1c is such a good measure of your average glucose over the past few months), fats, and other cellular molecules, altering their shape and function, triggering your immune system, and resulting in inflammation and auto-antibodies.

It may be belaboring the point, but it's important to recognize sugar as the colossal cognitive insult that it is. And the truth is that these are just *some* of the many pathways through which glucose and insulin dysregulation may lead to neurodegeneration.[7]

To make matters even worse, much of our sugar intake has been replaced by high-fructose corn syrup. Some have euphemized this sticky substance as "just like sugar," with the idea that standard sugar (sucrose) is something we are all used to, and is really not so bad (!). However, as dangerous as sugar is, fructose is actually worse for your health. As Prof. Richard Johnson of the University of Colorado has pointed out, there is a remarkable number of parallels between the brain's effects from fructose metabolism and early Alzheimer's disease, right down to the areas of the brain that are most affected.[8] And remember the AGEs, those bad actors that accelerate aging? Fructose produces them about ten times faster than glucose![9]

This may sound paradoxical to some: After all, sugar is processed into energy, our brains need *a lot* of energy, and too little is just as much of a problem as too much. But a human body operating on a healthy diet has *plenty* of glucose to burn. Supplementing with even more sugar is akin to already having gas in the tank of your car, and yet standing with the hood open, car running, and pouring gasoline all over the engine!

So, yes, sugar is the big bogeyman. But it's not the only one. Taking out a "big boss" is a huge first step, but I encourage you to imagine this as a John Wick–style thriller, where the assassins just keep coming, one after another, sequel after sequel. To ensure an ageless brain, you are going to have to beat all of them. And just like the characters made famous by Keanu Reeves, Jackie Chan, or Bruce Lee—martial arts masters who somehow survived these repeated onslaughts—you can do this. Thankfully, unlike those guys, you don't have to fight them all at once. Our lives are long and our bodies are resilient. You can take it slowly. Sugar is a great first battle to fight, but there are other brain-robbing insults, so let's get to work on them.

TOXIC AVENGERS

Another common insult that is now known to contribute to dementia is toxic exposure—not the sort of "bad guy falls into a vat of chemicals" sort of calamity that might come to mind, but rather the tiny exposures you are likely to face today, in your own home, during your commute, and in your place of work. As noted earlier, these toxins fall into three groups: inorganics (like air pollution and mercury), organic toxins (like anesthetic agents and glyphosate), and biotoxins (such as the mold-produced mycotoxins, trichothecenes, ochratoxin A, gliotoxin, zearalenone, and citrinin). We'll delve into each of these categories in chapter 11, but long before we get there it's important to point out that toxins, just like sugar, have a direct impact on brain aging and also accumulate as contributors of disease.

Our toxin exposure may be from the off-gassing of volatile organic compounds in new furniture, carpets, paint, and clothing. It includes the chemical components of fine particulate matter in the air you breathe when you go outside—in fact, one-third of particulate air pollution in the United States is from fires, which need not be close by to pollute our air. It includes the arsenic, lead, nitrates, chlorinated disinfection byproducts, uranium, and polyfluoroalkyl substances that are all too often found in municipal drinking water. It includes insecticides, pesticides, and herbicides. Exposures to these and other potential "dementogens"—and especially the long-term retention of these toxins in one's brains, bones, other organs, and blood—are increasingly being tied to the development of dementia.[10]

It is now clear that toxins like these are quite effective at altering the patterns of molecules, called methyl groups, that attach to DNA and modulate genetic behavior—the very sort of "epigenetic" dysregulation that many researchers are coming to believe is a key cause, if not the primary cause, of cellular aging.[11] Epigenetic changes are those that affect the readout of your DNA—in other words, which genes are turned on and off in each cell. Moreover, these modulations can be carried by sperm and eggs, and thus toxic exposure effects in one generation can be passed down to the next, and the next, and the next after that.[12] In an environment that didn't change much from one generation to the next, epigenetic inheritance may have primed offspring for survival, generations before they were born.[13] Childhood trauma, for instance, has been shown to be one of the most intergenerationally transferable experiences,[14] often materializing as anxiety and anxiety-like behaviors. This is perhaps because, in a very dangerous world, younglings, primed by the experiences of their forebears to be a little more risk-averse, stood a better chance at making it to adulthood and procreation. But in an environment that changes a lot—with all manner of new toxins to deal with in a rapidly industrializing world, for instance—an epigenome that is very good at "remembering" could be our undoing. The toxins our brains encounter today are bad enough without having to contend with those our grandparents and great-grandparents lived with as well.

There's little hope of avoiding all toxins in this world, but thankfully we have a dynamic system—we are constantly excreting, inactivating, and sequestering the many toxins to which we are exposed. So, it is a good idea for all of us to know what our toxic load is and ensure that we tip the balance away from further accumulation and damage.

WE CONTAIN MULTITUDES

If you were among the many millions of people who suffered from "brain fog" during a bout with COVID-19, you know how much a pathogen—in this case the novel coronavirus that caused a global epidemic—can impact your memory and ability to concentrate. What you might not realize is that these and other neurocognitive symptoms may be related to marked increases in biological aging that often impact people after bad bouts with this disease,[15] one of many examples in which viruses, bacteria, fungi, and parasites can "hijack" our bodies, affecting the major causes of aging.[16]

Once again, if pathogens, like the coronavirus that causes COVID, were only accelerating aging, that would be bad enough. But pathogens can more directly cause disease and stoke existing diseases as well. For instance, when individuals who were already suffering from dementia contracted COVID, it accelerated their structural and functional brain deterioration. In one study, a year after recovering from COVID, all patients had experienced significant increases in fatigue, depression, loss of attention and memory, and speech difficulties, among other issues. Brain scans showed that these patients were suffering from an increase in neuronal death, or brain cell die-off—far more than would have been expected under normal disease progression.[17]

Unfortunately, even without underlying neurodegeneration, COVID has the potential to cause massive damage to brain cells. This is why many young and seemingly healthy people were pushed into a state of acute neurodegenerative loss after just a single bout of the virus.[18]

There are many concerning parallels between COVID-19 and Alzheimer's. When the virus that causes COVID is cleared out, early enough, by the immune system, the long-term consequences tend to be minor. Unfortunately, in many cases the innate immune system goes through the roof, because the virus has a mechanism to camouflage itself inside our cells, delaying the initial, innate response.[19] When the immune system does finally get going, it's already overwhelmed, and the body is flooded with cytokines, the vital messenger proteins that signal the immune system to engage the enemy but which can cause the immune system to go into overdrive, attacking not just invading pathogens but healthy cells as well. This is known as a "cytokine storm." Similarly, with Alzheimer's, the adaptive immune system fails to clear the various insults—including, as we now know, many kinds of pathogens. But death does not come from a cytokine storm but rather from "cytokine drizzle," years and years of mild elevation of proinflammatory cytokines such as interleukin-1-beta, interleukin-6, interleukin-8, and tumor necrosis factor alpha.

Over the past few years it has become quite common for patients who have been treated for cognitive decline—and done well—to contract COVID-19, leading to a step back in cognition. There seem to be some cases as well, in which administration of a vaccine—although safe for many people—appears to activate the immune system and trigger a return to cognitive decline. This is a tremendous reflection of just how "on the edge" our brains can be, where healthy functioning can be offset by seemingly minor insults.

In stark contrast to the old view of brain aging, cognitive decline, and neurological disease as things that are inevitable and immutable, however, it's clear that these sorts of setbacks can often be set right. For instance, I recently spoke with a couple who had been on our ReCODE Protocol for just a few months. The husband, who has Alzheimer's at the mild cognitive impairment stage, was already beginning to improve.

The wife, who was following the protocol to be supportive and proactive about her own cognitive health, noted that the brain fog associated with her symptoms of long COVID had disappeared.

COVID, of course, is just one of many pathogens. A recent study of 1.7 million people found a strong correlation between hospital-treated infections (such as pneumonia, urinary tract infections, and surgical infections) and subsequent dementia, and the scary thing is that this doesn't happen immediately—the interval between infection and dementia averaged nine years.[20] Periodontal disease—a chronic infection that elicits a constant trickling of bacteria into the bloodstream—is also another ticking time bomb for Alzheimer's disease.[21] This all makes sense. Anything that activates the innate immune system, creating inflammation, is a potential contributor to brain aging and disease.

This major driver of cognitive decline is often ignored. Pathogens are not sought, not considered, and not treated by the vast majority of neurologists in their evaluations of patients with age-associated cognitive decline. Thankfully, this is beginning to change as the understanding of the various microbiomes—gut, oral, skin, and others—improves and doctors wake up to the fact that, given that there are more bacteria in the human body than actual human cells,[22] it simply makes sense that these tiny organisms collectively play important commensal roles, including as agents against other pathogens.[23] In fact, in classic "if you can't beat 'em, join 'em" fashion, we evolved to *need* bacteria: Without any bacteria in our bodies, we would be unlikely to last more than a few days.[24] One of the most important areas of research today is aimed at understanding the balance between how bacteria help us and how they can potentially hurt us.

Protozoa, worms, transmissible misfolded proteins known as prions, and especially fungi can also wreak havoc on a brain running at its max. We'll get into all that in chapter 12, but for now it suffices to note that the multitudinous minuscule creatures we carry along with us—from before we're born until well after we die—are important players in the long-term health of our brains.

ENERGETICS

Let's say you are pedaling a bicycle and trailer uphill. In the trailer are your family's groceries for the week, a tank of oxygen for an elderly relative, and a large bag of flower seeds that you're intending to sew in your front yard. You're cycling at maximum capacity, and for the moment, you're doing fine. You're still moving uphill and getting closer to your home. But let's say you accidentally drive over some weeds that are growing in the crack of the road and—just your luck—the plant in question was *Tribulus terrestris*, also known as goathead, devil's thorn, and puncture vine because its spiky, woody burrs seem uncannily drawn to bicycle tires. And now your back tire has started to lose air. It's not totally flat, yet, but there's still a mile to go before you get home, and you don't have the energy to compensate for the extra work it takes to ride a bike with a flat tire. Alas, it looks like you're going to need to leave something behind. Perhaps you can come back for it later.

So, what would you take out of the trailer? Yeah, me too. Hopefully nobody steals the big bag of flower seeds left on the side of the road, but that's definitely the right call. And now you've got *just enough* energy to get home. Your family will be fed. Your elderly relative will be able to breathe. And all it cost you was something that is nice to have but isn't mandatory for survival.

That's pretty much what's happening when we "hit a burr" in life, too. When we breathe in polluted air or are exposed to a nefarious chemical. When we pick up a malicious virus or bacterium. When we're insulted in myriad ways, life gets a little harder. And this is when the energetics that sustain the youthfulness of our brains become so very important.

There are four energetics that we will focus on throughout this book.

The first is sufficient blood flow. We need about 750 milliliters—that's the amount that would fill one wine bottle—each minute, with 250 milliliters traveling up each carotid artery (on the way to supplying the frontal lobes, temporal lobes, and parietal lobes) and 125 milliliters traveling up each vertebral artery (on the way to supplying the brainstem, cerebellum, and occipital lobes). It's well established that changes to vascular structure and function that happen with aging impair our blood flow, and thus, the youthful functionality of our brains.[25]

One of the main proteins in all that blood is hemoglobin, which is the substance in red blood cells that transports oxygen across the body with remarkable consistency throughout our lives. Sufficient oxygen saturation (SpO_2), preferably in the 96 to 98 percent range, is another energetic vital to brain functionality. But we know that oxygen utilization is hampered by aging as well,[26] another big knock for cognitive functioning. Sleep apnea, pulmonary disease, heart failure, air pollution, and sedentary lifestyle can all compromise this crucial part of the energetic supply.

In chapter 1, I noted that another energetic of great concern, our mitochondria, are built like finely tuned racecars, but the excellent performance we get from these organelles when we are young comes at a long-term cost: The size, volume, strength, and functionality of mitochondria decrease when we experience aging.[27]

The final energetic we will discuss in depth is raw fuel—the carbohydrates, fats, and proteins that come from our diets. Aging does a number here as well. It has long been known that, as we age, our metabolism declines; our bodies simply get worse at turning fuel into energy.[28]

If we accept the premise that all these energetic supplies have been primed from hundreds of millions of years of evolution to operate at maximum capacity, and that performance begins to falter with aging, then it is actually quite logical to assume that cognition would be one of the things we would expect to see fail as we get older. Something's got to go and, like that bag of flower seeds, the most logical thing to leave on the side of the road is that which is a "nice to have" but not a "need to have."

But what falls into the "nice to have" category? Well, the ability to learn new things, for one. We can get by for a long time on the stuff we already know.

What else? Memories of things that happened a long time ago. Or things that happened just yesterday but aren't likely to be super important for today. Or the ability to know whether something important happened yesterday or long ago.

This is what our brains are doing when we suffer from diminished energetics: jettisoning that which is unlikely to be immediately important in service to the instant task on hand, which is *always* our survival in this very moment. Sometimes the things that are abandoned are truly not important. And it's vital to remember that memories are not direct recollections of experiences, but rather a reinterpretation of those experiences based on other information—especially the shared recollections of others. So, if there's something to leave on the side of the road, it's those shared recollections. As a social species, we have evolved in such a way that others can help us remember important things if we forget.

This is why it's so common for us to forget things like what we had for lunch yesterday, what we wore to work or school the day before, or what television show we watched on the day before that. Most of these things aren't important to survival and, if they do turn out to be important, chances are that someone around us can help us recall. Moreover, these days, there's a good chance that we can pull a little computer from our pocket, check the photos, messages, and calendar items, and have enough contextual clues—if not direct evidence—to conjure back the thing we need to remember. We can thus very safely leave these items on the side of the road, as we can either go back for them or someone can help us retrieve them.

This is also why that pesky little demon who whispers into our ears—telling us not to worry about things like memory, focus, and mental flexibility—is so persuasive. Because it turns out that we can lose *a lot* of our executive functioning and memory storage and still get by. It's not usually until we have another flat bike tire or a broken axle on the trailer or a wobbly set of handlebars that we realize we're not going to make it home. And now we have to decide between the oxygen and the groceries! But, of course, these are not conscious decisions. Our brains will decide for us. And what is left in the trailer is not always what we will actually need.

WELL DESIGNED . . . FOR ANOTHER TIME

A while back I heard the inspiring story of Craig Humburg, a carpenter who was in his early fifties when he was diagnosed with ALS in June 2006. Craig was told by his doctors that the mean survival time for someone with his diagnosis was three to five years, but he didn't want to live those years in resignation of his terrible fate. So, even as his muscles began to atrophy and he developed some struggles with speaking, he went cycling five days a week, thousands of miles each year, on the roads around his home in northern Iowa. "It's my freedom," he said in 2014, years after he was originally projected to have been gone from this life.[29] Exercise helped Craig live far beyond his original prognosis.

Exercise is essential to fighting neurodegeneration—and moreover, as we'll discuss in depth later on, to *prevent* neurodegeneration and *all* forms of cognitive decline, including that which we have been led to believe is a "normal" part of aging. For the purpose of this analogy, I'd like to focus on Craig's very cool recumbent tricycle, which was outfitted with specially designed armrests and controls to help with shifting and breaking—a vehicle perfectly designed for his needs on the relatively flat roads of Cerro Gordo County, Iowa. Now, if we'd take that same vehicle on a rugged mountain trail, we'd find ourselves in dire straits. For journeys like that, the best bicycles are mountain bikes, built for durability, with suspension forks to help with shock absorption, large knobby tires to grip the dirt, wide handlebars to improve balance, and a big difference in gear size to provide the torque needed on rapidly varying terrain.

Both cycle types are well designed for the needs of their users, the products of many years of iterative specialization, but neither would be a good substitute for the other. And this is a good way to think about our genes, too.

Despite what we've often been told, our personal set of nucleotide sequences can't really be segregated into "good genes" or "bad genes." Rather, the genetic variants we carry are often specialized for one thing or another, which can make them beneficial or detrimental for what they're actually being used for, especially in the context of genes that evolved over many generations to do one very important thing but have been thrust into a modern world in which that task is no longer applicable for survival.

An excellent example of this comes in the form of genetic variants that are associated with a significantly increased risk of another neurodegenerative disease, multiple sclerosis (MS). In 2024, the members of an international team of scientists from twenty different universities and research institutions revealed that they had tracked the evolution of many of these variants back to a group of pastoralist peoples who lived across the Pontic Steppe region of Eastern Europe and Central Asia thousands of years ago.[30] At its most basic, MS is a heightened immune response. And for those ancient herders, the evolution of a stronger capacity to recognize and counterattack an invading pathogen would have been quite beneficial as a protection against zoonotic diseases—those that are passed from animals to humans, especially when two or more species are living in close proximity to one another. Ironically, given what we now know about pathogens, this immune response may have helped protect those herders from cognitive decline. And the fact that this extra protection would have caused an excess of inflammation, the accumulation of which would have deleterious effects arising around the ages of forty or fifty, wouldn't have mattered so much from an evolutionary perspective. By then, after all, these pastoralists would have passed on their genes and become parents and grandparents, whereas those without the antizoonotic genetics might easily have died before procreation, thus selecting for these MS risk genes.

But in today's world, where very few people live in such close proximity to herd animals, and those who do are protected by modern hygienic practices, vaccines, and other forms of medical care, that specialization is no longer helpful. In lieu of a constant barrage of pathogens to attack, the genetic variants that are associated with MS direct the immune system on a scorched-earth mission that destroys a person's supply of myelin, the protective layer that covers nerve fibers in the brain and spinal cord.

In one particularly instructive example, it was shown that antibodies directed against the Epstein-Barr virus's nuclear protein (EBNA1) cross-react with an adhesion molecule, Glial CAM, which is produced by the myelin-producing cells, the oligodendrocytes.[31] This is molecular mimicry, such that a person's immune response against the Epstein-Barr virus (EBV) generates autoantibodies against their own brain's white matter, causing or contributing to MS. What is not yet clear is why nearly all of us are exposed to EBV and yet very few develop MS: If you take 1,000 people at random, about 940 of them will have been exposed to EBV at some point, yet only 1 of those will develop MS. Why is that? It may have to do with genetics (at least in part related to the zoonotic protection described above), exposure specifics, or something else.

We see this again and again in genes that are often maligned as "bad," "faulty," or "defective." Doctors and researchers call these "pathogenic variants," but even that term denigrates a sequence of genetic code that likely served our ancestors in beneficial ways, even if those ways aren't exactly clear. Admittedly, though, this way of thinking does little to change the fact that, in the context of our modern world, these genes are akin to a cycle made for one purpose that is simply incompatible for the purpose we actually need it for, if we need it at all. The variants are, for all intents and purposes, part of a systemic network responding to insults that can contribute to cognitive decline and neurodegeneration.

Like the other categories of neurological insults we have discussed thus far, "fixing our genes" would not be enough to prevent cognitive decline. While the age of genetic manipulation may be exciting in many ways and offer hope for some people who carry genetic variants that are particularly poorly adapted to the modern world, the truth is that only a very small number of cases of neurodegenerative disease are caused by (deterministic) genes. In Alzheimer's, for instance, only about 5 percent represent familial Alzheimer's, and those are due to mutations in one of three genes: amyloid precursor protein (APP), Presenilin-1, or Presenilin-2. Thus, gene therapy is not a magic cure for diseases of the brain.

INSIDE THE TROPHIC CASES

We touched on the idea of trophic, or cell survival, factors in chapter 1. They are substances that provide something like bulletproof armor for our cells, enhancing survival even in the face of insults. The first of these was discovered back in the 1950s, when Rita Levi-Montalcini identified nerve growth factor (NGF). There are now known to be dozens and dozens of such factors, and they stimulate cell growth and survival, but when it comes to brain cell survival, there are three types of trophic support that are particularly important. The first are neurotrophic factors, which affect the brain, like NGF and brain-derived neurotrophic factor (BDNF), the latter of which is particularly important in Alzheimer's and is increased with exercise. The second are trophic nutrients, such as vitamin D. The third are trophic hormones, like estradiol, testosterone, and thyroid hormone, which help our brains communicate complex functions across vast physiological networks: when to sleep and wake, how to emotionally respond to different stimuli, how to process different kinds of foods, and much more. We'll delve into each of these in chapter 13, because optimizing supportive stimuli helps to avoid brain aging.

As we experience aging, our bodies struggle to process nutrients as efficiently as they once did. The fuel that leads to sufficient neurotrophic function and signaling might be as good as ever, and we might even *improve* upon the quality of that fuel by eating more sensibly as we grow older and wiser, but the machinery we need to process that food is simply not working as well as it once did. This is another insult that can be explained in the context of hundreds of millions of years of evolution. From the standpoint of species survival, the period of life that is most important for *any* sort of fuel to be effective is early on, through the point of reproduction and child rearing. So, trophic machinery that wears out in our forties or fifties has nonetheless still done its Darwinian job. But for those of us who wish to stick around after that point—with the audacity of desiring good cognitive health, no less—trophic changes are a pretty big problem, a challenge that evolution hasn't gotten around to fixing to accommodate our modern expectations of life.

WISH WE COULD TURN BACK TIME . . . BUT NOW WE'RE STRESSED OUT

In the Harry Potter series of books, the time-turner is a magical device that, as the name suggests, can be used to return to the past, thus changing the course of events in the present. This is, of course, not the only story in which time travel plays such an important role. Stories of people with the magical ability to travel into the past go back many hundreds of years, but outside of fiction, we have yet to develop the ability to turn back time, thus we cannot completely avoid danger. But evolution did provide what might just be the next best thing: It gave us cortisol, a hormone that has the almost-magical capacity to immediately raise the amount of glucose in our blood, resulting in a surge of energy that permits us to either confront or flee danger.

Cortisol is great—so long as it's used sparingly. But for brains that are already running like racecars, too many energy surges can blow out the machinery. Again, this isn't such a big problem in the context of species survival. As long as cortisol helps ensure short-term survival through reproduction, the long-term consequences don't matter so much. The problems we now face aren't only that we're getting more such surges as a result of living longer but we're also tapping into cortisol far more often than we ever needed to in our evolutionary past.

One example of this is the literal clamor of our modern world. If we took all human causes of noise out of the equation, the loudest sounds in the natural world would very rarely exceed 40 decibels, and in some parts of the world the loudest sounds would rarely exceed 20 decibels. By way of contrast, the noise level in a typical restaurant often hovers around 80 decibels, and at a rock concert, 90 to 120 decibels. Thus, during a *single meal* you could be exposed to sounds much louder than anything that most of your ancestors ever heard, and thus, anything they were evolved to handle. And, of course, restaurants are hardly the loudest environments any of us have ever been in. Sports stadiums, construction sites, airports, and concert halls all frequently expose us to noises that have been demonstrated to raise our cortisol to levels that would have been necessary only very sporadically during our evolutionary history.

Sound is far from the only stressor that raises this stress hormone. Research has shown that cortisol spikes can also be triggered by high-pressure work environments, financial stressors, relationship problems, and even something we often associate with relaxation: watching TV! Research has shown that television and movies can bump up cortisol, which can further aggravate the cortisol dysregulation that happens as a result of screen-induced sleep disruption and the sedentary nature of watching programs on a screen.

Our lives are filled with these sorts of stressors, and thus, filled with cortisol (and, by the way, several other stress hormones that were evolved for similar purposes and overused in the modern world). Aging is associated with changes to all these hormones,[32] so not only do stresses accumulate over time, but our bodies become less effective at producing the biochemicals we need to deal with that stress.

It seems logical that cognitive decline is considered a normal part of life at a certain age. By that time, after all, we have been overloading our neurological systems again and again, without so much as a story about surviving an encounter with a saber-toothed animal to show for it!

YES, IT'S A LOT

I understand if the implications of what I have just explained may feel a little overwhelming. Indeed, it seems impossible to do anything about them. I have heard many patients protest that while they *might* be able to stop consuming so much sugar, it seems crazy to think they can somehow turn back the clock on toxins, prevent pathogens, overcome genes designed for a different time, change their trophic trajectories, reduce stress—and on top of all that, that fight back against something so seemingly inevitable as aging!

My answer is that I have seen others do it. I have helped them through it. And every day we identify better and better ways to aid people as they take on this seemingly herculean task.

What's more, many of the people who have gone through the ReCODE Protocol have done so while experiencing *profound* cognitive decline. If they can do it, under those circumstances, the rest of us can, too, as a preventative pathway.

I'm not going to pretend that what I will describe in this book—the pathway to a healthy and fully functional brain at every stage of life—is easy. I will simply point out that it is much easier the earlier you begin, and with small steps and iterative optimization, it is highly effective and really not that hard.

I will also note that while it is true that overcoming any singular insult is very unlikely to entirely prevent or reverse cognitive decline, it is very rare to see anyone who does not experience some improvement in their cognitive state as a result of interventions that are aimed at any one insult. With each of these small successes, we build capacity to take another step, to attack another problem, and to keep moving toward the ultimate goal: a youthful, functional brain, for life.

3

WHAT IS POSSIBLE AT ONE HUNDRED AND BEYOND?

None are so old as those who have outlived enthusiasm.

—HENRY DAVID THOREAU

In 2024, a team of geneticists, neuroscientists, and supercomputing experts from Spain and the United States published a study that seemed perfectly designed to frighten millions of people around the world. Nearly seven million Americans, and 150 million people globally, have two copies of the variant of the apolipoprotein E (ApoE) gene known as ApoE4. The scientists' grave conclusion was that pretty much everyone in this genetic situation will develop Alzheimer's. On average, according to the study, the age of symptom onset was sixty-five years.[1]

But there's a difference between "pretty much everyone" and "everyone." And, at least so far, that difference seems to be represented by a man named Henry.

Like 7 million other Americans, and many more around the world, Henry is homozygous for ApoE4, meaning he has two identical variants (alleles) of this gene. But at the time of this writing, he still had excellent cognition and he wanted to keep it that way for many years to come.

I suppose I should mention that Henry is one hundred years old. And if Henry can make it to one hundred with two copies of ApoE4 and still have a youthful brain, I'm confident that we all can.

Can we go even further than that? Perhaps! In recent years there has been a lot of talk about whether it is possible—or even likely—to live well beyond the absolute upper limit for human longevity. Could we live, as some researchers have postulated, up to 150 years?[2] Or is it possible, as others have suggested,[3] that we could extend our lives to the maximum known longevity for other mammals—more than 200 years, which is the age that our distant cousin, the bowhead whale,[4] has managed to reach? Other vertebrates who are not quite so closely related, like the Greenland shark, may live as long as 500 years;[5] might we match their amazing lifespans? And while it is harder to envision, as we imagine ourselves to be so very different from other twigs on the tree of life, we are at our basal selves no more special than any of the other 9 million species of eukaryotic life-forms with which we share this planet. I'm referring to those life-forms with DNA in the form of chromosomes contained within a distinct nucleus. Other organisms that share this trait are known to live thousands of years[6] and, barring predation and catastrophic habitat changes, some are thought to be functionally immortal.[7] And so it is that some people have asked whether it might be possible that we could engineer humans in this way as well.

These are not silly questions nor merely fanciful ideas. For the record, I don't believe that any of these ideas is outside of the realm of *eventual* possibility. Given enough time and effort, humans might just figure out how to live longer than our ancestors could ever dream of. Humans have done the seemingly impossible before, after all, and there seems a good chance that we'll do so again, including, perhaps, when it comes to the years we expect to spend on this planet.

I hold three things to be true when it comes to lifespans that have, thus far, been unheard of.

First, just about everyone alive today has a chance of living to see a time in which our species blows past known limits for longevity. Except for those who are already devastatingly sick, we all have the capacity to benefit—right now—from the revolutionary and rapid movement toward that goal. We're seeing improvements to so-called normal cognition losses many people start to experience in their thirties and forties. The likelihood for such longevity will improve substantially, perhaps exponentially, based on how healthy we remain between now and then. Biological aging is one of the key factors, if not the capital consideration, for such long life.

Second, to have a revolutionary impact on most people's lifespans, we only have to get closer to our known potential. And to do that, we wouldn't have to change anything. Even as things stand, there is a tremendous gap between the average age at which we depart this plane of existence, which is about eighty years old in most developed nations, and the known maximum human lifespan of about 120 years.

Third, virtually no one wishes to arrive at any age without a sound mind. Indeed, I have lost count of the number of people who have told me that their greatest fear is living into old age without being able to think rationally, remember reasonably, and recognize loved ones. (If you have this fear as well, no matter how you might feel ethically about assisted suicide, you can certainly understand why it is such a common topic of discussion among those who have recently been diagnosed with Alzheimer's disease. The prospect of life without a sound mind is rightfully terrifying.)

So, yes, we can dream of 120, 150, or whatever age we wish. There's really no harm in that sort of dreaming. As they say, a goal is just a dream with a deadline. This is particularly true if your dream is guided by a rational understanding of the social and scientific trajectories that might actually permit such things to happen. But if we wish to get to *any* of those ages, and do so with a sound mind, we still must travel through that which we already know is possible. So, the number we will concern ourselves with in this chapter, and for the rest of this book, is one hundred. And the goal we will set is one in which we get to that age with absolutely no degradation in mental capacity—a brainspan of one hundred wonderful years.

What Is Possible at One Hundred and Beyond?

That alone would be world-changing! Remember, after all, that some degree of cognitive decline has long been assumed to be underway for most people when they are in their forties—and, as we discussed in chapter 1, the network of insufficiencies that drives this process begins with an accumulation of insults that happens *long* before that point in life. Insulin resistance, toxin accumulation, pathogen exposure, energetic diminishment, genetic risks and damage, trophic changes, and negative stress are all factors that have the capacity to begin priming our brains for cognitive decline when we are still in the womb! And that, according to much recent research, is likely when biological aging begins as well.[8] So, it's no small task to get to one hundred at maximum mental capacity. Indeed, people like Henry are still very rare, even when they don't carry multiple copies of an allele that puts them at greater risk for neurodegenerative disease. But we also know that brainspans like that are, in fact, possible. And there are more folks than Henry who have managed to do it.

You might think people like Henry would be cognitive unicorns—so rare that it's fair to ask whether they really do exist. That's what researchers in the Netherlands figured they'd discover when they gathered more than two hundred centenarians who had self-reported to be cognitively healthy. Now, you might already (and correctly) have noted that this is not at all a representative sample of people who have reached their hundredth years. Individuals who believe themselves to be cognitively healthy are, presumably, more likely to indeed be cognitively healthy. But remember that what we are concerned with is not the *average* experience at one hundred but rather a reasonable *potential* experience at one hundred, and what the research team learned from these people who believed themselves to be cognitively healthy is that those self-assessments were often right.

The researchers employed a test that is similar to the MoCA, albeit with a less catchy acronym. The Mini-Mental State Examination (MMSE) is a quick-to-administer assessment of memory, attention, and executive function, among other cognitive attributes. Like the MoCA, it is scored out of a total of 30 points, although it is not quite as sensitive as the MoCA. Generally speaking, a MMSE score of 25 or higher is associated with normal cognitive health at *any* age.[9] Some of the participants, it turned out, were unable to complete the test, not because of cognitive challenges but rather because their vision or hearing was diminished. Of the 151 individuals who did take the exam, though, about half scored a 26 or higher, which generally would be interpreted by a clinician to be an untroubling score even for a person many decades younger. About one in twenty of the test takers got a perfect 30 out of 30.[10] At one hundred years old and beyond, these individuals were as sound of mind as any human is expected to be at the height of their cognitive health!

But what does this study of relative outliers mean for the rest of us? What do such scores reflect about a person's day-to-day life? And can we really reach one hundred in perfect cognitive health?

I'll answer that last question first: Yes. And I want to share how you can do it.

COGNITIVE GUIDE STARS

When he began his near daily workouts at the age of sixty-one, Les Savino was already one of the oldest regulars at his gym in Hanover, Pennsylvania. Forty years later, his regular "lift day" at the gym included more than nine hundred reps on fifteen different machines. On aerobic days, the former US Army Air Corps pilot and manufacturing executive would do eight miles on a stationary bicycle and two miles on a treadmill.[11] "Most people when they get older give up," Les said in 2022. "You know, it's an effort to come in here every morning. That takes a lot of effort, and people don't want to extend that effort. They just want to sit in a chair and watch television."[12] But that wasn't what Les wanted to do in his so-called "golden years." And what struck many of the employees at the gym, as well as everyone else around him, was not just his ability to exercise for three hours each day at the age of one hundred, but his impeccable memory, astute wisdom, and acerbic wit.

Les, they observed, was not a man whose body and mind were moving in two different directions, but rather he was a person who was as cognitively healthy at one hundred as most people in their thirties and forties.

Howard Tucker was still a practicing neurologist at Saint Vincent Charity Medical Center in Cleveland, Ohio, when he reached his hundredth birthday, seeing patients and mentoring doctors young enough to be his great-grandchildren. Howard was in his late nineties when the COVID-19 pandemic hit, and although older people around the world were being discouraged from maintaining contact with the world around them, Howard says, "I had to work. I put on a mask. The hospital didn't tell me to stay home. They said everyone should come to work. In medicine we have a responsibility. If you take it seriously, you follow through."

When he wasn't at the hospital, he and his wife, who was ten years his junior, would read the newspaper together every morning over breakfast. Howard was known for his ability to quote everyone from the statesman Winston Churchill to the poet Dorothy Parker,[13] and he developed a thriving following on social media. What I have come to adore about Howard is that he wasn't just using his remarkable cognitive health for himself, but in service to so many others.

Sisters Ruth Sweedler and Shirley Hodes remained well connected to the details of their amazing lives, and the changes happening in the world around them, long after they surpassed their 100th birthdays. "My doctor loves to talk to me," the younger sister, Ruth, said in 2023. "He'd say, 'You're amazing.' And I'd say, 'Because I'm old?' And he'd say, 'No! Because you're sophisticated.'" Three years older than Ruth, Shirley bristled at the very *notion* of age. "I'm not that old!" she said shortly after her 106th birthday, noting that no one who retained a curious mind was ever *really* old. "Some people aren't interested in anybody but themselves," she said. "I was always so interested in hearing people's stories, backgrounds. They're full of surprises."[14]

People like this can serve as our guide stars, offering us the ability to track our own journeys against that which we know to be possible. And they aren't alone! There were about 750,000 centenarians in the world in 2024, and that number is expected to quintuple in the next thirty years.[15] And what this means is that while not everyone who reaches one hundred will do so in perfect cognitive health—indeed, these cases will continue to be outliers for some time—we will absolutely hear more stories of people like Henry, Les, Howard, Ruth, and Shirley.

I do realize that it can sometimes be difficult to identify with people we read about in books and news stories or see on television programs or social media videos. So, while I absolutely believe that we should seek out such stories and use them as inspiration, I also encourage people to look around their own families and communities for individuals who have gotten closest to the goal of a hundred-year brainspan. These will, in many cases, be people who reached their eighties or nineties without any apparent cognitive decline. The benefit of examining folks who are close to us is that we are more likely to see not just what is possible, but also to be aware of things these people might have done *better* to get even *further* down a path that leads to a hundred-year brainspan.

That's something I think about a lot these days.

SHARP TO THE END

Philip was ninety-two years old. Most of his friends had long since retired, and quite a few of them had passed away. But he had come from a family of immigrants who arrived in the United States in the early 1880s—part of a huge wave of arrivals from Norway, during that decade, who had helped populate much of the Midwest, contributing to an agricultural work ethic that the region continues to be known for. Those values remained deeply seated in Philip for his entire life.

In school and his early career, he had developed a reputation for identifying elegant solutions to applied physics challenges—a penchant for problem-solving that he took with him into civilian service following World War II, when he had worked on improving Allied missile technology. After the war, as the United States entered a period of industrial growth and commercialization that was often rooted in engineered solutions, Philip was responsible for the development of a variable-speed, ladder-type conveyor, the likes of which you'll find today carrying hamburger patties over a flame broiler in fast-food restaurants. Over the years, he'd launched multiple companies. And although his body held up quite well, relative to many others, as he surpassed his seventieth, eightieth, and ninetieth birthdays, it was really his mind that permitted him to continue to work. So, that's just what he did.

"If I can still contribute, why would I not work?" I recall him saying when I once asked him whether he would ever consider retirement.

I remember being constantly and tremendously impressed with his sharp mind, and I consulted with him about my own endeavors. Although he had no formal training in medicine or biological research, his observations and advice were quite insightful, even prescient, and I often found myself thinking about what he must have done right, on behalf of his brain, to be able to engage in such conversations with ease as he approached a century on this planet. Clearly, he had kept his brain active and engaged, always on the lookout for new problems to be solved; research tells us that is good for long-term brain health.[16] He was a charming guy and never lacked for people who wanted to be around him; we know from many studies that strong community connections are associated with longer brainspans, too.[17] He was an avid reader, and always had some new book with him. Because he had been fairly successful in business, he had everything he needed and could afford most of what he wanted, but he never engaged in excess. He ate a healthy diet,[18] didn't gain weight as he aged,[19] worked frequently in the yard and stayed fit, had regular checkups and screenings,[20] and maintained a healthy blood pressure throughout his life.[21] Research suggests that all these habits are helpful for preventing cognitive decline later in life, and, indeed, on his ninetieth birthday he looked like someone in his sixties.

Philip passed away after an illness shortly after he turned ninety-three, having remained mentally fit until the last month or two of his life—a tremendous example, I think, of the idea of "squaring the curve" of healthspans, which suggests that we can live full and healthy lives, right up to the end, and then move on quickly, rather than experiencing a slow and painful degeneration over many years or decades. If everybody could experience life in this way, it would have a massive impact on the healthcare system and our very notions of what life can and should look like. Philip was a great model for what is possible.

But after his death, I did spend quite some time considering the things he could have done *even better*. There was a fair amount of strife in his life. He was a World War II veteran, and I know he carried the weight of that experience with him. He had two broken marriages and some significant regrets; evidence suggests that divorce is a strong predictor of early mortality, especially for men.[22] And, like many people of his generation, he drank alcohol, although never to excess. For a time, moderate alcohol consumption was thought to be associated with a lower risk of cardiovascular disease, but more recent research that balances potential benefits to one system with consequences to others has suggested that there is really no amount of alcohol that is not deleterious to holistic health.[23]

It is a fool's game to play, to want to change the past, but I'll always wonder if this remarkable man, so mentally fit in his early nineties, could have remained that way for another decade if the decisions he had made over the course of his life had left him healthy enough in other ways to fight off the illness.

I would have given just about anything to find out. Philip was my father.

STAYING LONGER

Something on the order of 110 billion people have come and gone before us.[24] For now, it remains a good assumption that each of us will follow them, likely somewhere between 80 years (roughly the global average in developed nations) and 120 years, which borders upon the uppermost limit of known human longevity.[25]

Our children will likely miss us when we are gone. Our grandchildren and great-grandchildren, too. If we live a life that is good and honorable, we will be mourned by friends and family.

Maybe it will indeed come to pass that maximum human lifespans will increase. But if we do not increase our brainspans in kind, this will be a terrible crime against those we love, for they will mourn us all the sooner and for so much longer. For these reasons, I see the protection of cognitive health not only as a gift to our future selves but also as an obligation to those who love us, will love us, and will care for us.

It is thus incumbent upon each of us to set a course for a guide star, to identify someone who has done it as well as it seems it can possibly be done, and then resolve to try to do it just a little bit better.

If we do, our brains at one hundred will be able to do the following important things.

Retain Sharp Memory

A person with perfect cognition at one hundred years old is able to recall vivid details from the past, all the way back to the early years of their childhoods. As noted earlier, people developing Alzheimer's or age-related cognitive decline tend to lose the ability to form new memories while retaining memory for events earlier in life. Essentially, the brain is faced with insufficient power for its tasks and strategizes to retain what has been learned, which is a very successful strategy in most cases. I often meet new patients who have lost the ability to learn new information but function at a very high level with what they have learned over their lifetimes. Of course, the goal is to retain the new memory as well, but this repeated observation demonstrates how successful the brain's strategy is.

Learning something new is a bit like a surgical procedure: Many things have to go right, in the right sequence, to achieve the intended outcome. You have to focus, which is why people with attention deficit disorder often have difficulty learning. Your brain assigns a level of importance to the various incoming stimuli, which is why burning your hand on the stove is not something that is easily forgotten. You must have multiple forms of synaptic transmission (especially from acetylcholine, the "memory neurotransmitter" that is reduced in Alzheimer's); enough trophic support from neurotrophins such as BDNF, hormones such as estradiol, and nutrients such as vitamin D to form and maintain those synapses, a good supply of an intracellular signaler called cyclic AMP, enough thiamine, the ability to alter specific ion channels in order to modify neuronal excitability, functional mitochondria,[26] the ability to alter the architecture of neurons by constructing dendritic spines (tiny outpouchings on neurons where synapses occur), not too much inflammation, not too many amyloid-beta oligomers (these interact with complement C1q on synapses and prune the synapses),[27] not too much toxin exposure, and intact neuroanatomy (especially the hippocampus and related structures). I'm being willfully abstruse here, for I wish to emphasize how absolutely remarkable it is that our memory works so consistently when so much has to go right for that to happen!

Consider now what must have been going *right* in the mind of Catherine Walter, who was able to retain her memories at the age of 107 whenever she was asked about growing up in Chicago's Riverdale neighborhood during the onset of World War I. Nearly one hundred years after the war began, she could still vividly remember the soldiers who came to guard the bridge over the Little Calumet River near her home. "Every day they would walk uptown to get their groceries and they always walked by our house," she recalled in 2017. "They'd stop and visit with us children if we were out. All of the families took turns sending them treats."[28]

Centenarians with sharp memories not only remember a few events like this but many. Importantly, they can also remember more recent events and retain new information with minimal forgetfulness.

Engage in Complex Problem-Solving
Being able to recall lots of memories at the age of one hundred and beyond is a wondrous thing, but if we struggle to employ all that knowledge in the present day, it's of far less value. Yet we know it is possible for

people with hundred-year brainspans to engage in complex problem-solving tasks, analyze information, and devise effective solutions, thus showcasing their continued cognitive flexibility and adaptability. That's what Yuri Averbakh was still doing as he came upon his hundredth year of life—becoming the first-ever chess grandmaster ever to reach that age. The lifelong problem-solver, who had studied as an engineer and repaired tanks during World War II, was still trying to unlock the complex puzzles of his game at the age of one hundred.[29] At that age, he was still meeting with colleagues to share the ideas that would come into his mind. "Sometimes I analyze endgame positions," he said on his hundredth birthday in 2022. "I understand that in the computer age these analyses have no practical value, but this activity helps me to keep my mind sharp."[30]

Learn New Skills

At one hundred years old, a cognitively healthy brain should be capable of learning new skills, whether it's a language, a musical instrument, or a newly developed technology.

Even complex video games should be no problem for a person with a hundred-year brainspan. Indeed, when Kit Connell was gifted a handheld game console by her daughter for her ninety-sixth birthday, she easily learned how to use it and began amassing a large collection of games, which she was still playing on her hundredth birthday. Among her favorites, as it happens, was a game called Brain Age that was based on cognitive training activities developed by the neuroscientist Ryuta Kawashima, who was among the early and most outspoken advocates for "brain training" to combat cognitive decline. "I can't speak highly enough of this Nintendo," Kit said of her console in 2012. "It helps to keep my brain as active as possible in my old age. If there's any secret to a long life it's to think positive and keep your mind active." In fact, she used her first console so much that she wore it out and needed to get a new one.[31]

The ability to continue to learn new skills, such as Connell's capacity to learn new games, is demonstrative of healthy plasticity, meaning that even at one hundred years old the brain can still change structure and function to fulfill new objectives.

Maintain High-Level Reasoning
A healthy hundred-year-old brain will have the ability to think critically, make sound judgments, and engage in logical reasoning processes.

One of my favorite examples of a person who has retained all these cognitive qualities is Stanley Sacks, who began his law practice after serving in the US Army Air Forces during World War II and was still litigating personal injury cases more than seventy-five years later, making him the oldest practicing attorney in the United States at the time. "I'm just as interested as when I started," he said in 2023. "The jousting between attorneys . . . is a war in itself. It's the scrapping it out that I enjoy so much."[32]

To "scrap," as Stanley still enjoyed doing at the age of 101, requires flourishing executive functions, the mental processes that enable us to plan, organize, problem-solve, focus attention, regulate emotions, and control impulses to achieve goals. Those who master these functions are far more likely to enjoy social and professional success at all points in life—including, it is clear, many decades after the vast majority of people have retired from mentally demanding careers.

Sustain Emotional Connections
There is little I can think of that would be more fulfilling than still to be sharing my life with my beautiful and brilliant wife, Aida, when I am one hundred years old. And if I do make it that far, I'll most certainly have her to thank—as an integrative physician, she has taught me more about this field than anyone. A cognitively healthy brain at the hundred-year mark is one that can engage in meaningful relationships that provide emotional support and mental stimulation.

There is an increasing number of people who have become guide stars for this indicator of long and healthy brainspans, but a couple I adored learning about was Morrie and Betty Markoff, who reached one hundred after nearly eight decades of marriage, and "still haven't killed each other," Morrie joked in 2017. "Though we've tried a few times," Betty added. "We've had plenty of run-ins. Oh my God."[33]

Alas, very few people who make it to one hundred will do so with the love of their life (indeed, the odds right now hover somewhere around one-thousandth of 1 percent), but it is very hard to get to that point without strong emotional and social connections. In turn, these are connections that—as Morrie and Betty have shown—can, indeed, still be sustained by centenarians with cognitively healthy brains.

Exhibit Emotional Stability

A person with a hundred-year brainspan is no less able at a century to maintain emotional resilience, effectively managing the stresses around them and coping with life's challenges in a constructive manner. And given how much life will change around us, if we commit to sticking around for that long, this is a vital attribute of cognitive health.

A person who turned one hundred in 2025, for example, likely endured at least some of the impacts of a global depression during their childhood. They came out of that experience only to enter adulthood to find a world at war. They spent the remaining years of their early adulthood in a world changed by globalization, weapons of mass destruction, and terrorism. Around the age that many become grandparents, they watched as the world split under the influence of two superpowers, each with the capacity to destroy human life on Earth with little more than the push of a button. As their grandchildren began having children, they experienced a world suddenly transformed by an information revolution, an interconnectedness that brought with it the opportunity to spread misinformation at an unfathomable scale. And as they approached their one hundredth birthday, they were subjected to a pandemic that claimed the lives of millions of people and created profound suffering and social isolation. And these, of course, are just some of the geopolitical stressors they faced. Along the way they suffered the deaths of their own grandparents, parents, siblings, and friends. The communities where they lived were transformed many times over, and the prevailing mores shifted.

A healthy brain, at one hundred years old, has sustained all this and yet it can still handle more. It can, in fact, offer compassion and empathy for others who are struggling with the difficulties of life. That is what Hedda Bolgar was doing well past her one hundredth birthday. The psychologist was still seeing patients four days a week, training new therapists, and traveling across the country to give lectures, including an influential address to an educational affiliate of the American Psychoanalytic Association in which she spoke about the need to reject dogma and embrace flexibility in service to patients whose lived experiences will be very different from generation to generation.

To be clear: It is not that a cognitively healthy person does not mourn and suffer as the world changes. Hedda certainly did. When her husband died, she recalled, "it was really, for many years, the end of the world. My mourning was endless. It seemed endless, until one day I decided I was alive." With that, she rededicated herself to others. "Ultimately what really interests me is to see people change and have better lives," she said in 2011.[34]

Demonstrate Mental and Physical Creativity

Individuals who are cognitively healthy at one hundred remain creative, generating novel ideas, expressing artistic talents, and thinking outside the box in various aspects of life. This seems to be the easiest-to-imagine positive attribute of a person with a hundred-year brainspan, perhaps because many people tend to envision centenarians as necessarily sedentary beings, and many forms of creativity can be consummated without tremendous mobility. Indeed, it is not hard to find stories of people who, at one hundred years and beyond, are still painting, whittling, knitting, and writing.

But it should also be said that the ability to apply physical coordination, balance, and motor skills in service to creative pursuits, like sports and dance, is not simply an attribute of a healthy body, but also a healthy mind, and I'm not sure I've seen a better example of this than Shirley Goodman, a tap dancer who became a social media darling in her late nineties when her family began posting videos of the "Dancing Nana" online. She was still shim-shamming, scuffling, shuffling, and stamping along on her one hundredth birthday. "I would advise people, if they like music at all, to keep it in their lives and don't just sit home in a rocking chair," she said in 2023.[35]

I highly recommend pulling up a video of Shirley and, if you do, take note of the fact that she is not just going through the motions of the dance steps she has been doing since her vaudeville father taught her to tap when she was eight years old. Watch her eyes and facial expressions as she connects with the people around her. Watch her transition as the accompanying band shifts from song to song. Watch her mouth as she sings along. This is a brain operating on many levels at once and, when you recognize it in this way, it's an even more inspiring thing to see.

MAKING THE POSSIBLE PROBABLE

We know it is possible to reach the age of one hundred while retaining crystal clear memories of our youth, as Catherine Walter did. We know that people of that age can still solve complex problems, as Yuri Averbakh was still doing as he rounded the corner on a century of life. It is possible to learn new skills like Kit Connell, maintain high-level reasoning like Stanley Sacks, sustain emotional connections like Morrie and Betty Markoff, exhibit emotional stability like Hedda Bolgar, and demonstrate mental and physical creativity like Shirley Goodman.

A person who exhibits *any* of these attributes at one hundred would be an inspiration to many, as all these individuals are to me. A centenarian who exhibits *all* these attributes, though, has accomplished the goal we have set forth in this book. For all intents and purposes, they have enjoyed an ageless brain—and the important implication here is that if you get to one hundred in this way, then you have also gotten to ninety in this way. And if you get to ninety with full cognitive health, then you have also reached eighty in the same manner, and so on. Yes, we know it's possible to restore cognitive health, but this truly is the path that anyone would prefer—no degradation at any age, for any length of time. That is a true hundred-year brainspan.

We know this is possible. Now, it's time to set about making it probable. That will take work. But I have yet to encounter anyone who cannot at least *improve* their cognitive health, at any age and any stage of neurodegeneration, and there is thus no reason why we cannot proactively protect ourselves from cognitive decline—including, in most cases, preventing it altogether through our sixties, seventies, eighties, and nineties. Even with so-called bad genes, like Henry. Even with some unfortunate life circumstances, like Hedda Bolgar, or a few unhealthy habits, like my father.

I understand that a hundred-year brainspan might seem difficult to imagine right now, especially if you know that your own health is not perfect. It's worth noting, though, that nobody's health is perfect!

That includes me.

As we will explore in much greater depth later in this book, sleep is absolutely integral to cognitive health. But I've spent a lot of my career staying up late into the night, taking care of patients, conferring with researchers around the world, and even sitting alone at my desk trying to work out the complex challenges of battling neurodegenerative diseases. We'll also talk a lot about stress and good eating habits. I've failed in these regards, at times, and as a result there have been a few points in my life in which my blood pressure has been less than ideal, which has well-known neurological implications.[36]

So, I recognize I haven't done some of the things that my own cognitive guide star, my father, Philip, did correctly. I'm trying to do better with those things, and succeeding in some ways that he did not. In addition, I am looking to other guide stars, like some of the centenarians you have met in this chapter, and informing my choices with the available science. Of course, none of us knows what will come next year, next week, or even in the next minute of our lives, but I am in my seventies now, and I feel quite cognitively fit and have every reason to believe that

What Is Possible at One Hundred and Beyond?

I have the capacity to stay that way. So, you see, the goal here is not perfection in all things. That is a terrible expectation to set and, moreover, it would be a hypocritical one for *me* to set for you.

Instead, I'd like to invite you to take a step toward your personal cognitive guide stars. When you succeed in that step, take another. I have seen in the lives of my patients, and affirmed through my own experience, that each small action toward better cognitive health begets the capacity to take another action, and the next step is usually a bit easier than the last. Alas, it's never effortless, but it does begin to feel *natural*—especially as the rewards become palpable—as if we are moving toward a place we are supposed to be.

4

DYING OF PROFIT

> The further up the ladder you are, the more acceptable it is to lack vision; if you want to be a complete idiot, you'd better be an expert...
>
> —R. F. LOEB

My nephew, Bryan, is a remarkable guy who writes symphony music and modern music, plays many instruments, and has released several albums as part of a band.

He turned out pretty well, I'd say. But when he was a boy of only three, his family lived next door to an older boy named Paulie, who taught Bryan to cuss like a sailor. This resulted in Bryan spending many, many hours alone in his room. One day, my mother—who adored Bryan and dearly wanted for him to break his horrible habit—sat down to reason with him.

"Bryan, now just stop for a moment and think about all of the many hours you have spent in your room, simply because you have been repeating those nasty words you learned from Paulie," she said.

She could see the beginnings of recognition flashing across Bryan's face, and she knew she was on the right track. "Now, does that teach you anything?"

Bryan squinted his eyes with what she thought was an important new insight. Then, suddenly, he blurted out, "Yeah! That fuckin' Paulie!"

For Bryan at age three, the insight wasn't quite there yet—understandable at age three, but, disappointingly, we see the same lack of insight—with disastrous patient impact—when we allow profit to drive medicine. You can see this clearly in three recent documentaries, jaw-droppingly good films that showcase remarkable medical progress but also the stifling of the pace of this progress—and even the obfuscation of results by those who richly benefit from the status quo. While each of these three films portrayed the struggle to identify better therapeutics for three different diseases, I was surprised that I came away from each with a similar feeling.

The first was *The Quiet Epidemic*, which is about patients who developed chronic Lyme disease (CLD). The "chronic" part of this is important, because the mainstream understanding of Lyme is that of an acute infection that, while dangerous and even deadly in some cases, is usually over quickly, without long-lasting consequences. The poignant stories in this film, however, show the lives of people who have long suffered after contracting this disease—at times confining them to wheelchairs, in others requiring heart transplants, and in some cases ending their lives. It also meticulously documents the scans, blood tests, and electrophysiological tests that document the chronic damages that can be wrought by this illness.

Richard Horowitz, arguably the world's leading expert on CLD, has written books and developed protocols that are helping many people to survive and thrive after contracting chronic Lyme. But despite the many patients, symptoms, testing, scans, admittedly difficult treatment, and experiences of a growing number of physicians who, like Horowitz, believe Lyme to be a potentially chronic disease, the insurance companies *deny the very existence of this disease*! (This is like a climate-change denier delivering a lecture on a 140-degree day.) What's more, mainstream medicine has been pressured to follow suit, and sadly it has yielded and done just that. Indeed, the movie makes a compelling case that the health insurance companies realized quite quickly that chronic Lyme was going to be expensive, so the best way to save money was simply to deny its very existence. Of course, the insurance companies then need to get some "medical experts" on their side but, just as we've seen with Alzheimer's, there's always someone who can be bought. (Which is exactly what a group of Lyme disease patients alleged in a long-running suit against the Infectious Diseases Society of America, which the plaintiffs allege conspired with insurers to create arbitrary guidelines that said Lyme can be treated in virtually all cases with twenty-eight days of antibiotics, forcing some patients to pay hundreds of thousands of dollars for care they believe should have been covered by their medical insurance. In November 2023 the US Court of Appeals upheld the dismissal of their case, holding that "statements about chronic Lyme disease constitute nonactionable medical opinions.[1]) This impressively effective (and offensive) practice of denial in the face of overwhelming evidence raises the question of whether the same companies could deny the existence of other diseases, like cancer and cardiovascular disease. I have little doubt that they would do just that were it possible, but since those diseases predate the reimbursing companies, the insurers use other tricks, such as "preexisting conditions," to deny as many claims as feasible, and keep the cash flowing into the corporations.

The other two documentaries tell very similar stories through the lens of different diseases. In *Living Proof*, a new and effective treatment for MS is denied even while the patients themselves represent the titular proof. In *Memories for Life—Reversing Alzheimer's*, several of the patients with whom I've consulted or corresponded over the years, such as Lucy, Sally, Deborah, and Frank, all discuss their own documented improvement; yet again, there is utter denial and pushback from the insurance policymakers.

Is there any other system in which you hand over your money, then ask for some of it back, but the decision to return it is unilaterally in the hands of the group that stands to gain from denying you? What a conflict of interest! This is not insurance. It's a scam.

Other disease experts, who are often paid as consultants by the various corporate insurers and health providers who stand to benefit from their opinions, argue that they are simply looking for "evidence-based" medicine. Or they say that, sure, a different intervention might look promising, but larger trials are needed. These can be legitimate concerns. So, how do you tell the difference between desire for progress and desire for profit? It's easy: When there is a desire for progress, there are invited debates and discussions, with a mutual goal of improving outcomes. When the goal is profit, the debates are squelched so that no opposing view is heard or sanctioned. That is clearly what has happened in the Alzheimer's market . . . I mean field . . . no, actually I do mean market.

The basis for this subterfuge and chicanery is elegantly discussed in a wonderful book by Robert Lustig titled *Metabolical*, featuring dastardly appearances by Big Food, Big Pharma, and Big Healthcare. The high seas in the days of Bluebeard the pirate were far less treacherous than what these modern-day health-plundering buckaneers have created: a food system that features ultraprocessed "food products" that have led to an epidemic of metabolic disease (and thus compromises brain health and aging); a pharma system that pushes hyperexpensive drugs that often offer little (or nothing, when it comes to improved brain aging and avoidance of neurodegeneration); and a healthcare system that supports these ineffective drugs over prevention and treatment of the very causes of illness (and will almost certainly deny your claims for reimbursement to reduce your brain aging). The result: a short healthspan, a short brainspan, and a short lifespan. The United States is not even in the top thirty in the world in longevity, and yet pays far, far more than any other country for its healthcare!

Make no mistake: The United States has made it clear that, when it comes to health, it prioritizes corporate incomes over patient outcomes. If we are to stay sharp for a century of living, we'll need to do far better than what is currently being pushed on us by "the Bigs." That is exactly what I hope this book will do for you—help you to make a turn away from the current model of sickness and into a protocol for long-term brain health and performance.

WHY OUTLIERS SCARE THE SYSTEM

Cognitive guide stars like Catherine, Yuri, Kit, Stanley, Morrie, Betty, Hedda, and Shirley aren't just outliers, they are the *outlyingest* of outliers. They're the exceptionally small percent of people who didn't just make it to the age of one hundred but also made it to that point in life in astoundingly good cognitive health. There's very good reason to believe that stories like theirs will become increasingly common in the coming decades.

Today's centenarians are people who were born in the first few decades of the 1900s. Global demographic trends strongly suggest that someone who was born in the middle of that century is even more likely to reach the age of one hundred, and someone who was born in the latter decades of that century has an even better chance, and someone who was born around the turn of this century has an even better chance. Indeed, by some estimates, half of the people who have been born since the year 2000 will live to one hundred years.[2] Other researchers who have sought to model potential changes in longevity through the end of this century are less bullish but nonetheless see coming improvements in lifespans across the globe. For instance, in 2017 a team from Imperial College London concluded that the tremendous gains in life expectancy across the world would inevitably slow in coming decades, but in the thirty-five nations they evaluated, not a single one was likely to lose any of the gains made toward improved life expectancy.[3] In virtually every place on Earth, people are living longer and are expected to keep doing so.

Increased lifespan, however, does not necessarily confer an increased brainspan. Most people I speak to seem to recognize this. Consider a conversation I recently had with a patient named Prakash, who was just fifty-six years old when he was diagnosed with moderately severe cognitive decline. Like so many other patients that our teams have worked with, Prakash's symptoms have since disappeared. He has stopped blanking on the names of colleagues as he used to, he no longer forgets meetings and social engagements as he once did, and he's not losing his train of thought as he did on a daily basis just a few years ago. As a result of these improvements, the depression and anxiety he was experiencing following his diagnosis have also abated. What's more, because a major part of his turnaround came as a result of lifestyle decisions including more exercise, a better diet, and improved sleep, it's not just his brain that is functioning better—his entire body is as well. So, in my mind, Prakash has every reason to look optimistically toward many healthy decades to come. And yet when I made an offhand joke about starting a savings account for the purchase of a hundred birthday candles, Prakash looked at me as though I had told him he should be saving up for a journey to the center of the Earth.

"Oh, doctor," he protested. "Don't curse me like that!"

"What do you mean?" I asked.

"Please don't think I'm not grateful for what has happened in the past year," he said. "I am healthier than I have been in so long, and my mind is clearer than ever. But I've met a few people who have reached that age, and . . ."

He was quiet for a moment, as if trying to figure out how to say something in the most delicate way possible.

"And it doesn't look like a good time?" I asked with a chuckle.

He laughed as well. "That's a good way to put it," he said. "I guess I understand that it's possible, but it doesn't seem likely. So, I think maybe eighty-five or ninety would be nice—if my body and brain both hold up. That's decades longer than I thought I was going to get."

"You don't think it's probable that you'll live longer than that with a healthy body and a healthy brain?" I asked.

"Possible, sure," he said. "Anything is possible. But probable? I really don't think so. Not for me. Not for anyone, I think."

"That's the way things used to be—possible but not probable," I said. "But Prakash, if I told you that it's both possible *and* probable that you can reach your hundredth birthday with great cognitive health, would you believe it?"

He thought for a moment.

"I guess I'd have to," he said. "When we first met, I really didn't think it was possible that my mind could feel this way again. But I think I would wonder the same thing that I wondered when we started working together."

"Which is?"

"If this is really possible—or probable, as you say—why doesn't everybody know about it?"

I've been asking that same question for a very long time. First, it was in regard to the vast improvements we have seen as we've treated patients with Alzheimer's.[4] More recently, I've been wondering why everyone doesn't know that virtually *all* age-related cognitive decline is preventable. Why on Earth would it be so hard to get these stories into the public consciousness?

The answer, as you've already guessed, is a single word: profit.

PREVENTATIVE PROFITEERING

Let's pretend for a moment that you've been given an option to invest in a new company, called Iasie, at an opening price of $26.50 per share. And let's say, for the sake of this exercise, that this is a company you *really* believe in. It does something no other is doing. It's led by very smart people who have a long history of success in endeavors just like this one. You've pored over the prospectus. You've done the calculations. And you're very confident that, come this time next year, each share will be worth no less than $37.60. That's roughly a 42 percent increase in value, about four times the average US stock market return over the past half century. Not a bad investment, if it works out.

Now, let's say that you currently have every dollar in your investment portfolio tied up in just one company, we'll call it Uoy. (For the record, this is a very dumb investment strategy, but for the sake of this example we'll pretend your initial capital came from an eccentric grandparent who stipulated that to keep the principal—and everything that grew from it—you had to always invest in just one stock.) Lo and behold, it's mostly worked out. Year to year, you've stayed just ahead of the market index. Lately, though, you've noticed that this stock is not performing as it once did, and you can see the company is doomed to failure. Its stock price is virtually guaranteed to diminish in value, year after year, until it's absolutely worthless.

How fast would you move your money? Pretty quickly, right?

That's essentially what leaders from a real company, called Eisai, were hoping people would do in 2023. The US Food and Drug Administration granted accelerated approval for its new Alzheimer's drug, lecanemab (sold under the brand name Leqembi). Eisai set a price of $26,500 for a year's prescription. But by the company's calculations, the "societal value" of the drug was $37,500 per year. And in a world in which Alzheimer's patients have long been told they truly have no hope anyway, what would patients have to lose? Ergo, the Japanese pharmaceutical company's leadership team had concluded, the drug was worth every penny!

You might have noticed that our fictional companies, Iasie and Uoy, are just the backward spellings of Eisai and You. And, indeed, a lot of people pointed out that Eisai's calculations seemed a little bit backward—if not altogether ridiculous—because Leqembi simply doesn't work very well for many people, and it can have dangerous side effects.

Markku Kurkinen, a professor at the Center for Molecular Medicine and Genetics at Wayne State University in Michigan, acknowledged that, in two clinical trials, Leqembi did appear to reduce amyloid in the brain—a major marker of Alzheimer's disease and longtime target of those trying to stop this disease. But, Kurkinen warned, the drug doesn't seem to slow cognitive decline in women, who have a twofold increased risk of Alzheimer's over men. What's more, Kurkinen pointed out, it doesn't slow cognitive decline in those with the genetic markers that make Alzheimer's most likely—or in the highest risk group, those with ApoE 4/4 (roughly seven million Americans), in whom the drug is associated with a *worse outcome than placebo*. Also, the drug doesn't *stop* Alzheimer's in anyone; at best, it modestly slows the progression of the disease. These and other results, Kurkinen wrote in 2024, "make me wonder if the approval of lecanemab was the worst decision of the FDA up till now." A close second-worst, he quipped, was the approval two years earlier of another "blockbuster" Alzheimer's drug, aducanumab, which is sold under the brand name Aduhelm, with an even more audacious launch price of $56,000 per year.[5]

Now, here's the thing: The companies that brought Leqembi and Aduhelm to market are absolutely right in one regard: Alzheimer's patients have long been told there's no hope anyway. Studies I've conducted with some colleagues have shown that's simply not true,[6] but let's say for the sake of argument it is accurate. For people at the very end of the line—who either have $26,500 per year to spend or an insurance carrier willing to fork over that amount (or part of it)—perhaps there really is no harm in hoping for a miracle from a drug that seems to work in some people, sometimes, in some ways, albeit only to forestall the inevitable.

But that sort of "desperate times call for desperate measures" approach is a far cry from what was being proposed by the Alzheimer's Association (whose largest single sponsor for many years happened to be Eisai) at the group's annual international conference in Amsterdam, Netherlands, in July 2023. At the conference, which was attended by about eleven thousand doctors and scientists, a twenty-person panel of peers recommended a radical new way of diagnosing Alzheimer's disease. Regardless of whether someone is actually experiencing any symptoms of cognitive decline, the panel suggested, anyone who tests positive for elevated levels of amyloid (or tau, which is another protein that has long been tied to Alzheimer's disease) would be diagnosed with "stage 1" Alzheimer's, thus making them medically eligible to be coerced into taking a prescription—Leqembi, Aduhelm, or the similar Kisunla (donanemab)—despite the fact that there was no evidence these drugs would reduce the risk of dementia among nonsymptomatic people.[7]

In the wake of the 2023 conference, *Los Angeles Times* investigative reporter Melody Petersen, a longtime pharmaceutical industry watchdog, began wondering who was on the panel that was making this revolutionary recommendation.

The answer, Petersen discovered, was at least seven people who were employed by pharmaceutical and medical testing companies, and seven more who received money from those companies for consulting or research. Additionally, at least four "outside advisors" to the panel were executives from Eisai or another company that sells or is developing similar drugs. Petersen was also interested in understanding why the panel's proposal included the National Institute on Aging in its title. Did the panel have the backing of the US government's leading center for research on Alzheimer's disease and other dementias? After she asked, the institute quickly took its name off the panel's proposal.

SO THINLY VEILED

I wish I could say I was shocked by any of this. But I am not.

Quite a few years back, I was invited to a meeting of Alzheimer's specialists hosted by a large pharmaceutical manufacturer. "We're gathering experts from around the world," a representative for the gathering explained to me. "We're going to meet in San Francisco, we're going to talk about each expert's vision for the future, and we want you to share your opinions."

I suppose I was a bit naive back then. At the time, I thought, "Well, this is neat! And right up my alley! Because when we talk about the future, we set a course to change the world!"

But when I arrived, along with about twenty other scientists, an immediate change in plans was announced. "We do want you to tell us about the future," a man who was a representative of the pharmaceutical company told us. "But, for each of you, your vision must include our two new Alzheimer's drug candidates."

What they were really looking for was a method to dole out their drugs as quickly and profitably as possible. Their main concern—the raison d'être for the meeting they had called—was that they felt there weren't enough neurologists to prescribe the drugs as quickly as they'd like. After all, once the drugs were approved, the clamor to obtain prescriptions would overwhelm the neurologists, and so one suggested strategy was to have all primary care physicians prescribe them. This would allow many more prescriptions for the new drugs to be written.

I was a bit surprised that no one else there seemed to be miffed by having been brought to the meeting under false pretenses. The meeting was not about vision or the future; it was about how to sell more drugs more quickly.

As it turns out, though, both drugs ended up failing in clinical trials. So, the future the company executives demanded we envision didn't come to pass. What a pity, right?

TESTING? YES. DRUGS? NOT YET!

The shame of the Big Pharma–stacked panel's proposal is that, although it was just another thinly veiled attempt to sell more drugs, it did indeed get something right.

For years, we've waited to treat diseases of the brain until after the symptoms of cognitive decline get bad enough that patients stop listening to the demons who conspire to tell them that everything is "normal" and come to a physician for help. It is unfortunate but often true that by this time, they've leapt over mild cognitive impairment and are suffering from full-blown dementia. It was long thought that there was simply nothing that could be done at that point to help. In more recent years some doctors have decided that there were *some* things that *could* be done for *some* people *sometimes*, but only to slow down the progression of symptoms, never to stop it altogether. We now know that it is possible to slow, stop, and even reverse these symptoms, but even the most optimistic physicians acknowledge that the worse they are, the harder it is to stop them. So, the key is to identify cognitive wellness-to-disease transitions long before people become symptomatic and, better yet, avoid these points of inflection altogether through scientifically validated preventative measures.

In this way of thinking, it's rational that anyone carrying a key biomarker for Alzheimer's should, in fact, be diagnosed as "stage 1." This is akin in many ways to how physicians have lately become much more aggressive in identifying prediabetes and pre-prediabetes (insulin resistance prior to prediabetes) to unlock a toolbox of interventions that respond to an actual medical diagnosis. It's true that one of those interventions is drugs, but there is so much that can be done before needing a prescription. For example, while the drug known as metformin used to be prescribed almost exclusively for people with metabolic dysfunction meeting the diagnostic criteria for type 2 diabetes, it's now sometimes prescribed for individuals whose doctors can see will be "on their way" to full diabetes unless some kind of intervention is made. However, there are many other approaches, including a plant-rich ketogenic diet, exercise, and supplementation (bergamot, berberine, and others). In fact, it is relatively common for us to see people who have begun a precision-medicine protocol for cognitive decline who end up no longer needing antidiabetes drugs, antihypertensives, or cholesterol medications.

I have no doubt that the executives of Eisai and other Alzheimer's drug companies would be thrilled if they could get even a small percentage of people who have been diagnosed with "stage 1" Alzheimer's, under the proposed new diagnostic criteria, to take their drugs. At the prices these companies charge, after all, it doesn't take too many patients to make a mint, especially if you can get them on those drugs for a few decades instead of a few years. But the big difference is that metformin has been prescribed for more than eighty years. Its well-known side effects are minor. And even without insurance, it costs just pennies a dose. Yet physicians have been reluctant to prescribe this drug for prediabetes. That's at least in part because, whenever it's possible to do so, it's always preferable to prevent disease with lifestyle changes and other more physiological interventions. In the case of diabetes, that includes blood-sugar monitoring, dietary restrictions, and exercise.

These are the same sorts of interventions that are at the beginning of the protocol we use to treat cognitive decline at every stage. Widespread and biochemically comprehensive diagnostic testing—the sort that the Alzheimer's Association conference panel suggested should be done to identify so-called stage 1 cases—would be a huge boon to our protocol. Our approach suggests that there are many, many things that can be done to help slow, stop, and reverse cognitive impairment long before we slide into dementia, and long before we need to resort to a powerful, body-altering drug with hemorrhagic side effects to keep that slide from happening.

I do want to be clear: I'm not antidrug, I am pro whatever gets the best outcomes, and I believe that the future cures will involve a combination of personalized, precision-medicine protocols and targeted pharmaceuticals. But the idea of limiting treatment to a single, minimally effective drug when there are published superior results is a decision based on income, not outcome.

I have prescribed pharmaceuticals to patients in the past and will continue to support doing so when the drugs are part of the most effective intervention. How could anyone consider anything other than offering the most effective intervention?

What would clearly be preferable in the case of so-called stage 1 Alzheimer's, just as it is for diabetes, are interventions that don't have potentially horrific side effects, like nausea, changes in blood pressure, flulike symptoms, dizziness, headaches, visual changes, worsening confusion, swelling of the brain, microhemorrhages in the brain, brain atrophy, and in rare cases, death. Those, by the way, are just the side effects of Leqembi that we know about. We are almost certain to discover many more side effects if this drug—which has only been tested in a few limited clinical trials, and prescribed for people who are only thought to have a few years left to live—gains traction as a first-order intervention for people who will take it for a much longer period of time.

If and when that happens, I'm quite confident that we will look back on this period of time in the same light that we view other pharmaceutical debacles of our recent past.

A BILL TO PAY

As an ex–pharmaceutical executive once said, "In my former world we would say, 'Where there's a will there's a way—and when there are more people to bill, there are more people to pay.'" If that little rhyme sounds like something you might have heard before—and if the Big Pharma–backed plan to open the market for often ineffective Alzheimer's drugs to more and more people makes you feel a little bit queasy—it might be because this is a story based on a very well-worn script.

The rise and fall of Purdue Pharma is a good example. The notorious drug company was cofounded by John Purdue Gray, who had found success selling a tonic made with glycerine, sherry wine, flower extracts, phosphoric acid, and a few other chemicals. In the days before federal laws regulated what claims could be made about nostrums like Gray's, the bottle's label suggested the potion would stimulate appetites, aid in digestion, increase the absorption of vitamins, and promote better nutrition. If it had done all that, it would probably still be around a hundred years later. It didn't, but the company built on these sorts of claims still was. By then, though, Purdue Pharma had moved into the Wild West of morphine sales, and in 1984 it began selling a version of that opiate that was designed to release itself slowly into a person's body. Purdue Pharma released a seemingly improved version, which it called OxyContin, that was approved by the FDA in 1996, quickly becoming a darling drug for pain management—even though no long-term studies of its efficacy and side effects had been conducted at that point.

We now know that Purdue Pharma was permitted to help write the FDA's official review of its drug with the aid of a medical officer who just happened to land a cushy job at the company later. And we know that Purdue worked very hard to market the drug as a first line of defense against pain. And that Purdue lied about the drug's addictiveness. And that doctors who prescribed the most drugs were often rewarded with trips to exotic locations and well-paid spots on a speaker's circuit. And that the family that was at the helm of Purdue, the Sacklers, hid billions in offshore accounts, declared bankruptcy on US debts, agreed to a payout intended to protect them from further judicial action, and in the end remained "extraordinarily, mind-bogglingly, teeth-grindingly wealthy."[8]

If this had been a singular tale, it would be horrific enough. But the same playbook that Purdue Pharma used has been copied and refined time and time again to sell drugs that barely—and often fraudulently—passed muster with federal regulators. Those companies then used those approvals to gain a foothold with doctors who would spread the gospel of the drug to patients who weren't at all like those who were in the studies upon which the approvals were based. One such example is Insys Therapeutics, which sold a highly addictive fentanyl spray that was intended to treat pain in cancer patients. The company encouraged physicians to prescribe off-label to patients with myriad other ailments, offering a sham speakers' program that paid big bucks to the doctors who wrote the most prescriptions.[9]

Hundreds of thousands of people in the United States have died from opioid overdoses in the past two decades, up to eight times higher today than it was before OxyContin was approved by the FDA. Heroin accounts for many of those deaths, but up to 80 percent of heroin users started their death spiral into opioid addiction with a doctor's prescription for drugs like those made and marketed by companies like Purdue Pharma and Insys Therapeutics.[10]

Today, it's plain to see that the cozy relationships between FDA officials, opioid companies, and many physicians were as fraught with conflict as can be. Yet many of the same mistakes are being made all over again with Alzheimer's drugs. Thankfully, this calamity is unlikely to result in an epidemic of addiction, but there will be costs, not the least of which is that these companies are working very hard to perpetuate a false narrative about the diseases they are purportedly trying to stop—namely, that the only things that "work," even a little bit, are the drugs they sell. That's hogwash. But with every dollar we spend, every minute we waste, and every opportunity we miss to treat patients in a way that actually works, we are further delaying progress toward a world in which neurodegeneration has been eradicated.

That's fine for Big Pharma, an industry that remains fat and happy so long as we are sick and taking drugs chronically. So, you see, this industry is not just marketing drugs that work barely better than a placebo; it's also marketing the idea that these are only the *latest* drugs, which suggests that others will come along at some point that are just a little bit better still. For this is an industry that understands that if you believe that there will *eventually* be a drug that solves this problem, you are less likely to take other actions today that prevent that problem to begin with—and that makes you a more likely (and perhaps longer-term) customer down the road.

So, when people like Prakash ask me, "Why doesn't everybody know about your successful published results?," I tell them that our medical system was set up a long time ago to treat sick people. Today there's a lot of money to be made doing just that. We are just beginning to figure out how to alter the system so that the driving force is keeping people well to begin with—but I do think we'll get there. The physiology of wellness versus illness is extensive and complicated, but with larger data sets for each person, AI analytics, and personalized, precision-medicine protocols, health and healthcare will improve.

That's the truth. I really do believe we'll get there. But those who wish to get there sooner need to know that they're up against a bevy of insults that adversely affect our cognitive health, the demons of denial that whisper in our ears, the evolutionary programming that prioritizes robust early health over longevity, physicians who think age-related decline is inevitable, and a lack of enough cognitive guide stars to make it feel as though a hundred-year brainspan is not something reasonable to consider. They're also up against a powerful industry that wants to make drugs the first choice, the second choice, and the *only* choice for treatment.

TIME AFTER TIME

While Albert Einstein's theory of general relativity, and concepts like wormholes and closed timelike curves, suggest that time travel could be feasible under certain conditions, the question of whether we could *change* the past to impact our present remains speculative, at best. I mention this because I find myself repeatedly telling patients that there's truly nothing that can be done about that which is in our past. What has been insulted has been insulted. What's been done to us, and even what we have done unto ourselves, is done. Alas, there is likely no going back to fix things, no matter how many DeLoreans we soup up with nuclear-powered "flux capacitors."[11]

We also cannot do anything about the history that got us to a place where profit often takes precedence over what is practical, let alone what is principled, when it comes to human health and well-being. Since the establishment of the biomedical model as the gold standard for medical training in the early 1900s, healthcare systems across the globe have been almost entirely focused on developing or acquiring medications with which to treat diseases, but only once those diseases have symptomatically materialized. We can't go back and change the trajectory of global medical enterprise, no matter how many police boxes we might power with the energy released by collapsing stars.[12]

Here I am, waxing historical, pop-cultural, and whimsical, but the serious fact is that the past is gone and the future, while uncertain, is contingent upon the present. So *now* is what we must work with. And yes, it's true, right now there is a lot working against anyone who wishes to completely prevent cognitive decline and thus head off nearly any chance of sliding into neurodegeneration and dementia. This is why we need to come prepared with every weapon in our arsenal.

Perhaps the most powerful weapon you will have, in this lifelong fight, will be the ability to articulate *why* you're in the fight to begin with.

5

IDENTIFY YOUR WHY

> The secret of man's being is not only to live but to have something to live for.
>
> —FYODOR DOSTOYEVSKY

It might seem obvious that anyone, and indeed everyone, would *want* to have a hundred-year brainspan. But if there is a truism in the field of public health behavior studies, it is this: What is obviously preferable in terms of long-term health is not always, or even often, what individuals take steps to pursue in their individual lives.

Perhaps the best example is the way many of us choose to eat. Decades of research has almost universally established that humans who eat plant-based diets live longer and healthier lives, particularly as a result of a substantial reduction in blocked arteries, thus a lower risk of heart disease, thus a lower risk of heart attacks and strokes. In one meta-analysis, scientists and clinicians at the University of Copenhagen examined the results of thirty randomized trials published between 1982 and 2022 to show the impact of various diets on age, body mass index, and health status. The data were quite clear: Vegetarian and vegan diets were associated with a substantial reduction in ApoB scores

(ApoB is a much better predictor of cardiovascular disease than simply checking cholesterol),[1] almost as if people who ate plant-based diets were taking a low dose of statins, but without any of the side effects associated with those drugs, such as an increased risk of diabetes.[2]

This wasn't shocking. Most people know very well that eating more plants is one of the key ways to improve a number of parameters of health, and many people can personally attest to the validity of research showing the palpable, nearly immediate effects of replacing processed grains, meat, and dairy with vegetables in their diets.[3] And yet the US Centers for Disease Control and Prevention has found that only about one in ten American adults is even eating the recommended daily serving of vegetables,[4] let alone making vegetables, fruits, beans, legumes, and nuts the dominant elements of their diets.

I do not at all wish to dismiss the myriad socioeconomic obstacles that stand in the way of healthy eating for many individuals, but the percentage of people in the United States who live in so-called food deserts, which are areas that lack sufficient access to fresh food, has declined in recent years. Today, about 95 percent of Americans live in census tracts that researchers consider to have reasonable access to grocery stores, neighborhood markets, or farmer's markets.[5] And while it is clear that in many cases fresh foods can be more expensive, which is another barrier that shouldn't be dismissed, research demonstrates that when it comes to foods that actually make people feel full and satiated, it can be cheaper to purchase fresh foods.[6] So, for many people, the choice between healthy, less healthy, and decidedly unhealthy diets is just that—a choice. And most of us do indeed *choose* to eat more sugar, more trans and saturated fats, and more ultraprocessed foods like packaged snacks and popular breakfast cereals.

All of this is to say that the challenge of eating healthier isn't really a question of insufficient science. It's a question of behavioral psychology—particularly the contrast between what we know in our minds to be good versus the actions we choose to take in our lives that we understand to be bad for us. So, before we move on to talking about the importance of nourishment for brain health—let alone all the other factors that go into ensuring a healthy brain for life—we have to talk about the psychology of changing personal behaviors for the better, because clearly it's not as simple as saying, "Well, that's both intuitive and supported by science, so I will do it."

First, though, I need to make an important disclosure: In the field of human psychology, I'm way out of my depth. Problem-solving? That's where I typically focus my energy. But logic and math seem to disappear when emotions and psychology take over.

"It's not your *intelligence* quotient that I'm talking about," my irreproachably patient wife has told me more than once when she feels that I am failing to understand something important to her, our family, or people we are close to. "It's your *emotional* quotient." Alas, she is right. I am admittedly somewhat challenged in that area, and still working on it.

So, in this chapter, as we focus on the psychological habits that can help improve our cognition right away and thus make a hundred-year brainspan likely, it would be wrong for me to present myself as an expert, for I am not a psychologist, my bedside manner as a physician is imperfect, and I have found myself quite frustrated by my own occasional inability to connect what I know to be good for me with the decisions I make in my own life. Instead, I am going to describe for you the psychological qualities that other experts have found in scientific studies pertaining to healthy decision-making, and the habits I have observed in my most successful patients. After all, these are the individuals who have managed to overcome the very big challenges that arise from cognitive decline, and they are also the people who have vanquished the peddlers of hopelessness who seem to be conspiring to keep us from taking action to protect our brains and ensure performance for life. If these patients can do it when already suffering from some level—and often a substantial level—of cognitive decline, just imagine what is possible for someone who hasn't yet begun that slide!

WHERE THERE'S A WILL

By the late 2010s, we had seen hundreds of people reverse their cognitive decline, and others remain stable for years, with our ReCODE Protocol. We had also begun to notice patterns of patient behavior that corresponded with the best outcomes. With good compliance to the protocol, cognitive decline was reversed in most patients, and the earlier treatment was started, the more complete the reversals typically were. "Let's keep optimizing," I often say to my patients, because that is what has led to the greatest sustained improvements. Many will improve for several years and then either plateau or undergo a mild secondary decline. These people almost always turn out to have new or previously unidentified contributors to decline, but when those are addressed, and the patient adheres to the adjusted protocol, their improvement resumes. To be clear, human brain physiology is complicated, and no one can do everything perfectly. But perhaps that's a good sign, because I know a lot of people who have done it *imperfectly* and still substantially improved their cognitive health!

Still, let's make no bones about this: The protocol is challenging for many people. And in almost all cases in which a patient's recovery was less pronounced or nonexistent, we found that the individuals had difficulty complying with our guidance. What's more, these are not onetime or short-term changes but adjustments that are part of a virtuous cycle of improving wellness. Thus, by definition, this process is never quite complete and will never cease requiring a tweak here, a change there, a new intervention here, another level of dedication there.

Because I've seen many people struggle, I must admit that when I first met Bruce I was concerned. The man standing before me was suffering from a level of cognitive impairment that bordered on the first stages of dementia. Even for someone who appeared healthy in every other way, Bruce had a diagnosis that indicated a network of insults and resultant insufficiencies that would be complicated to identify and overcome. But Bruce didn't appear healthy in every other way. Much to the contrary. He was easily one hundred pounds overweight. His skin was red and patchy—often a telltale indicator of inflammation across the body. Even though he had been sitting for several minutes before I walked in to meet him, he was breathing heavily. And although he was just fifty years old, he looked to me to be a decade older than that. So it was that my biases got the best of me, and as I drew a map in my mind of what success might look like for this individual, I imagined a line that has been sinking precipitously and that might smooth itself out. In other words, my gut told me we may slow down his impairment but probably couldn't stop it, let alone turn back the damage that had been done.

This wasn't the mindset I should have had. And almost from the moment we started talking, Bruce taught me why I shouldn't underestimate *anyone's* ability to make life-altering changes to reclaim their brain. Indeed, Bruce was so much more than he appeared.

"I've read your book," he told me as we shook hands. "Honestly, it took me a long time because it's harder to focus these days, but I was determined. I read a page, took some notes, took a break, and then got back to it, and I know that if we're going to fix what's gone wrong it's going to take a lot of work. I'm ready for that."

I was impressed by his pluck. But I also remember thinking that sentiments like these are so much easier said than done. I opened my mouth to say something to that effect, then thought better of it—because why would I undercut any motivation he was already feeling?

"OK then," I said. "Let's get started."

The connection between obesity and neurodegenerative diseases is well studied. One meta-analysis of thirteen studies that all ran for multiple years and each included at least one thousand subjects concluded that obesity in midlife, like that which Bruce was dealing with, increases the risk of Alzheimer's and dementia by nearly one hundred percent.[7] This hearkens back to what I mentioned earlier about insulin resistance—indeed, anything that reduces brain energetics must be addressed for best outcomes.

In addition, body fat is associated with systemic inflammation and thus may contribute to dementia, making the challenge of change even harder.[8]

One of the first orders of business would be getting Bruce's weight down and helping him become "metabolically flexible," able to utilize both ketones and glucose for energy, as quickly as it was safe and sustainable to do so. That would mean pairing him with a ReCODE-trained nutritionist. Bruce had also already done some investigation into bariatric surgery, so we spoke about the prospect that a procedure such as the Roux-en-Y gastric bypass, adjustable gastric band, and sleeve gastrectomy might lead to a reduction in further neurological damage.[9]

In whatever way he approached the challenge, if Bruce was successful in addressing his weight, it was likely going to have a big impact on his level of inflammation,[10] which is tested by checking the level of C-reactive protein circulating in the bloodstream. This is part of a simple blood test and should be about 0.3 milligrams per deciliter, and Bruce's level of 3.8 milligrams per deciliter affirmed that inflammation was indeed as pervasive throughout his body as I had suspected. Mitigating obesity generally results in lower inflammation levels because excessive fat often comes along with an excess of macronutrients that are stored in fat tissue, stimulating the release of proteins, peptides, cytokines, and other inflammatory mediators that move through the blood, infiltrating every system of the body.[11] Since inflammation also maintains a well-known relationship to cognitive impairment,[12] weight loss would essentially offer a benefit against multiple insults that are known to contribute to cognitive decline.

But obesity alone was not responsible for all of Bruce's inflammation, so we also discussed how important it would be to heal his gut, manage stress,[13] get better sleep,[14] stop drinking alcohol,[15] and consider the potential benefits of switching out the sodas he regularly drank for an anti-inflammatory green tea.[16] We further discussed the many medications that are often used to treat chronic inflammation, including nonsteroidal anti-inflammatory drugs like ibuprofen (which can come with some substantial side effects, especially when taken long term), corticosteroids like prednisone (which are immune suppressants that are also safest when only used for a short period of time), and biologics like adalimumab (which also suppress the immune system, albeit in a more targeted manner, but can nonetheless increase the risk of infection). These side effects make such drugs inappropriate to address the inflammation in patients with cognitive decline. However, the great news is that virtually everyone can reduce inflammation simply by normalizing their physiology, and we have many ways in which to do this.

First, we identify the source of inflammation—which for Bruce was likely to be largely caused by excess weight but could be exacerbated by a leaky gut, an undiagnosed chronic infection, gingivitis or periodontitis, chronic sinusitis, or other issues. Second, we treat the causes of the inflammation, whether it is to heal the gut lining or improve the oral microbiome or treat a tick-borne infection such as Lyme disease. Third, we resolve the inflammation using resolvins—omega-3-related compounds that were discovered by Harvard professor Charles Serhan and weaken innate immunity responses, enhance the body's capacity to resolve inflammation, and protect the organs.[17] Fourth, we minimize ongoing inflammation using curcumin, omega-3 fatty acids such as the DHA and EPA found in salmon and other fatty fish, the promising anti-inflammatory traditional herbal remedy known as cat's claw,[18] and other known anti-inflammatories like ginger or pregnenolone.[19]

Obesity and chronic inflammation can make it hard to exercise, which in turn makes it difficult to mitigate cardiopulmonary challenges like Bruce's strained breathing. But since poor pulmonary function is associated with accelerated dementia progression[20] and exercise has an inverse relationship to neurodegeneration,[21] it was clear that we needed to prioritize movement, so we also discussed connecting him with a personal trainer who specializes in healthy weight loss.

Finally, we broached the subject of aging. I suspected, and Bruce agreed, that he was likely experiencing aging much faster than most people. At the time, there were several tests that showed promise for measuring biological aging, such as epigenetic biological clocks and vascular elasticity testing, and we spoke about the correlation between these markers and neurodegenerative diseases.[22] Today, these are used routinely to determine biological age, and were instrumental in demonstrating the ability to reduce biological age[23] in some cases by simply using diet and lifestyle interventions that are quite similar to the principles that undergird our protocol for reversing cognitive decline.[24] Results like these are one of the reasons I am so optimistic about our ability to reduce the biological age of the brain, thus enhancing brainspans. An improved diet, more exercise, better sleep, less obesity, less inflammation, and better pulmonary health would all likely contribute to Bruce's state and pace of aging, but we also discussed strategies for boosting his levels of nicotinamide adenine dinucleotide, or NAD+, which plays a central role in aging and the development of disease.[25]

But even if we were successful in each of these areas, I told Bruce, there were still other insults we would likely need to identify and address.

"That's a lot," he said.

"It is," I said, feeling as though I was beginning to hear the sounds of hopelessness that so often echoed in my patients' voices when they learned just how hard it was. "So, I understand if you're feeling overwhelmed. If you think it's too much—"

"Oh no," he interrupted. "It is a lot. But what's actually overwhelming is imagining a future where my daughter doesn't have a father. Or maybe even worse, a future where I'm still here in body but my mind is gone."

"Oh," I said. "How old is your daughter?"

"She's four years old, doctor," he said. "I know I got a late start at parenting, but I don't want that to mean that I don't actually *get* to be a dad. I can do this. I can do all of this."

He paused for a moment and looked directly at me. There were tears welling in his eyes.

"I'm *going* to do this," he said.

Identify Your Why

It's often said that "where there's a will there's a way." Patients like Bruce have taught me that this is a very incomplete way of thinking, for we cannot simply summon willpower from the ether. The ability to control one's actions, emotions, and urges—to align what we *know* we need to do with our *actual* actions—requires us to borrow strength from everything around us, and especially from the causes, reasons, or purposes that drive our desire to have a healthier brain right now and to keep it healthy at eighty, ninety, and one hundred years or beyond. Thus, a better way of phrasing that old aphorism is "where there's a will there's a why" or, even more sequentially appropriate, "where there's a why there's a will."

THERE'S A WHY

Up until I started really focusing on the psychological qualities of my most successful patients, I had operated under the assumption that the best way to make day-to-day decisions is to align those choices with a long-term goal—to begin with the end in mind, as the author Stephen Covey wrote in his famous book, *The 7 Habits of Highly Effective People*.

For instance, if your objective is to retire with no less than five million dollars in the bank (and you didn't happen to already have some manner of millions of dollars sitting in an investment account) then it would be critical to align your day-to-day financial decisions with that ultimate ambition. That might mean accepting promotions that will require more time in the office, avoiding frivolous expenditures, or chasing professional opportunities across the country or overseas. These can be hard decisions to make, but if the question you ask yourself is "Will this choice get me closer to my long-term objective?" then these decisions often make themselves.

Perhaps money is not your end goal. Maybe, instead, your long-term objective is to spend as much time with your family as possible. If that's the case, then it might be very reasonable to decline to do work that pays better but requires more time, to decide against spending heaps of money on a vacation that your family doesn't need but very much wants, or to remain as geographically close to your loved ones as possible even if it means sacrificing some professional opportunity elsewhere in the world.

These are not mutually exclusive examples. For most people, most of the time, a well-defined, long-term objective may encompass many different attributes of a well-lived life, and focusing on one does not necessarily mean giving up on other successes. Oprah Winfrey, for instance, has written that when she was a child in rural Mississippi in the 1950s and early 1960s, the opportunities she could see for herself were limited. "You could teach in a segregated school. Or be a maid. A cook. A dishwasher. A servant," she wrote in 2009. From an early age, Oprah aligned her decisions to actions she knew she could take, and roles she knew she could fill. "I wanted to be a teacher. And to be known for inspiring my students to be more than they thought they could be."[26] That was something she could achieve in a meaningful way even if society didn't change dramatically in the years to come. When she was offered a job reading the news for a local radio station in Nashville, Tennessee, Oprah interpreted this potential journalistic role as a form of teaching—in alignment with her goal—and decided to accept the opportunity. This set her on a path to a television show that was watched around the world, a famous book club, a leadership academy for girls in South Africa, and authorship of books on subjects such as achieving happiness among other endeavors that are quite easily recognizable as aligned with her long-defined purpose of being an inspiring teacher. Today, of course, she is known as one of the most influential people in the world, having taught many millions of people, albeit not in the way she'd originally conceived.

Beginning with the end in mind offers clarity, directionality, and resilience, and I am still a firm believer in the power of this way of aligning the present to an intended future. However, when it comes to achieving the goal of brainspans of one hundred years or more, I have recently come to realize that this may be an incomplete framework, and I think this is especially true for individuals who are not *already* suffering

Identify Your Why

from cognitive impairment. When you finally stop listening to the deceitful demons and recognize that your brain is crying out for help, it's a bit easier to feel motivated to do something. Indeed, among my patients, the ones who realize "I am going to die if I don't do this" are often the ones who best adhere to the stringent requirements of the ReCODE Protocol up to and beyond the six-month mark, which is typically the point at which improvements become obvious. When everything seems just fine, though, it's easy for an end goal to feel like something you can take a circuitous route to achieve.

Thus, I've learned from patients like Bruce, it's not enough to know *what* you want. You need to know *why* you want it.

To explain, let's assume that our end goal, as informed by our cognitive guide stars, is to have brains that retain sharp memories, engage in complex problem-solving, learn new skills, maintain high-level reasoning, sustain emotional connections, exhibit emotional stability, and demonstrate mental and physical creativity—over the course of a long lifetime. These are absolutely fantastic "ends." But the work it takes to achieve these ends is not easy. It takes sustained efforts—consistently good day-to-day decision-making—over many decades. Unfortunately, it is an exceptionally well-established fact that people often struggle to do what they *know* they should do, right now, even if they understand that it will help them achieve a goal that they would like to realize far down the road. It's not that we can't conceptualize the ways in which good food, great sleep, habitual exercise, and other actions will affect us in the long term; it's just that the end alone is seldom enough to overcome the sacrifices that exist "in the now."

In the year that followed Bruce's diagnosis, as he battled to bring his weight down, address his inflammation, improve his cardiopulmonary health, and decrease his rate of biological aging, he fought against "in the now" thinking by constantly keeping his "why"—his daughter, Brianna—front and center in his decision-making.

"For the first few months I have to admit that I thought often about food," he later explained, "but the plant-rich ketogenic diet was wonderful, and began to improve my energy level very quickly." However, old habits die hard, and he added, "Even knowing that lowering my weight was one of the most important steps I could take to get my old brain back, I was constantly telling myself that I deserved this, or I really needed that, or that I could always double down tomorrow if I cheated on my really strict diet today. So, every day—so many times a day that I really couldn't even count them all—I told myself 'I need to be here for Brianna.'"

I think it's important to note that this didn't magically eliminate Bruce's cravings. But the thought of three possible futures for Brianna—one without a father, one with a father who was cognitively incapable of caring for her, and one with a father who was physically and mentally present for her—gave him a little more resolve to do the things he needed to do to overcome the myriad insults that had resulted in his cognitive deterioration. As his weight fell, his inflammation substantially subsided. As that happened, he was able to exercise more. That contributed to better oxygenation, which allowed him to do even more exercise—which, along with his improved nutrition, helped him drop even more weight. And while Bruce didn't end up doing any of the biological aging tests that were just starting to become available when he began this journey, he repeatedly attested that he was *feeling* younger, as if he'd slipped back into the body he'd had several decades earlier.

Slowly at first, and then at an increased pace, his memory and focus began to return, reflecting a pathway I've heard about again and again from patients and their loved ones. First, they notice that the decline has slowed. Next, they see that the decline has actually stopped. Third, they start noticing small positive movements, often increased engagement with the family or friends. Finally, they notice major improvements such as the return of vocabulary, interactions, navigation, and planning.

That was Bruce's trajectory, too. "From the start I knew it wasn't going to be easy, but when things were difficult, I focused on Brianna, and I thought I was doing this all for her," he said. "But it sort of turned out that she was doing it for me. If I didn't have her to think about when things got tough, I wouldn't have made it. I'd probably be dead, or I'd be in a situation that I personally think is worse than dead."

Not surprisingly, it turns out that among my most successful patients, children and other family members are a very common "why" that drives the day-to-day decision-making people must engage in if they wish to improve their cognition in the near future and substantially extend their healthy brainspans in the long term, too.

DON'T WHY ALONE

The nineteenth century saw great advances in neurology, with the identification of Parkinson's disease in 1817, multiple sclerosis in 1838, ALS in 1869, and Huntington's disease in 1872. Not long afterward, in 1906, a clinical psychiatrist and neuroanatomist named Alois Alzheimer reported on "a peculiar severe disease process of the cerebral cortex" in a presentation to a group of colleagues in Tübingen, Germany, and "Alzheimer's disease" would soon be added to medical texts around the world.[27] Along with these discoveries came the flourishing of several great institutes of neurological research, including at the Salpêtrière Hospital in Paris and the National Hospital for Neurology and Neurosurgery in London's Queen Square.

During this time, neurology and psychiatry were taught and practiced as one, referred to as neuropsychiatry, and this was the state of the field when Sigmund Freud began his training under Jean-Martin Charcot, the world's leading neuropsychiatric researcher and instructor.

However, as Freud's work on psychoanalysis began to gain a foothold with other researchers and practitioners, and as neuropathological studies revealed alterations in some diseases (such as stroke) but not others (such as neurosis), there was a fin de siècle split, with neurology focusing on the brain and psychiatry focusing on the mind.[28] That split has largely remained to this day.

More and more, however, neurology is being identified in psychiatric diseases, such as the recent work on energetic insufficiency in psychiatric diseases such as schizophrenia, from Christopher Palmer of

Harvard Medical School, Shebani Sethi from the Stanford University School of Medicine, and Nichola Norwitz from Oxford University.[29] In doing so, they have added compelling evidence to the work that has been done in the borderlands between neurology and psychiatry, creating the field of behavioral neurology. One of the early leaders of that field was an American neurologist named D. Frank Benson, a pioneer in brain imaging who had established a world-class program at UCLA by the time I arrived as a beginning assistant professor in 1989. Frank was a kind, dignified, and insightful neurologist who, among many other accomplishments, described new types of aphasia—difficulties with language caused by brain damage. He also described what has come to be known as Benson's syndrome—posterior cortical atrophy (PCA)—which he and his colleagues first identified in 1988 in five patients who were all experiencing progressive visual difficulties. As the formal name implies, this condition affects areas in the posterior part of the brain—mainly the parietal lobes (which run vertically above the ear) and the occipital lobes (which form the most posterior part of the brain) in areas responsible for spatial perception, complex visual processing, and calculation. The symptoms are quite different from Alzheimer's because memory is typically not affected initially and because it tends to begin in younger people. Nonetheless, it has turned out to be a distinct presentation of Alzheimer's, and between 5 and 10 percent of Alzheimer's begins as PCA.[30]

The mainstream assumption about PCA echoes what we often hear about Alzheimer's. As the Mayo Clinic notes: "There are no treatments to cure or slow the progression of posterior cortical atrophy."[31] However, I noted years ago that the nonamnestic presentations of Alzheimer's (those in which memory loss is not the earliest and dominant problem), such as PCA and primary progressive aphasia (PPA), in which patients suffer language impairment very early on, are often associated with toxins and pathogens.[32] Therefore, when brain health coach Kerry Mills Rutland contacted me about a person with PCA, a woman named Eve, I encouraged her to work carefully with Eve's physician to identify and address the many toxins and pathogens to which Eve may have been exposed. This isn't an easy process, particularly not for someone who is already suffering from the symptoms of a neurodegenerative disease. Fortunately, Eve had the support of her husband, Eric, who not only helped facilitate his wife's treatment but recognized the opportunity to minimize his own brain aging. Together, they became each other's "why."

Eve and Eric have been involved with Kerry for the past three years, continuing to optimize and address each contributor. Sure enough, it did turn out that Eve had both a pathogen—*Bartonella*, which is often acquired through a tick bite—and specific mycotoxins, which are produced by some mold species, such as *Stachybotrys* (the toxic black mold) and *Aspergillus*. As part of their adoption of the ReCODE Protocol, they consulted with Dr. Neil Nathan, a leading expert in the treatment of both tick-borne and mycotoxin-related disease.

Thankfully, Eve did not follow the course of decline that most other PCA patients follow, and, in fact, has improved remarkably. When she started treatment, she was unable to read or use a computer; both of those skills have since returned. In addition, her MRI volumetrics have improved dramatically: Her parietal lobe volume has increased from less than the first percentile to the twenty-second percentile, her occipital lobe from the eleventh to the twenty-fifth percentile, and her hippocampus from the sixth to the thirty-second percentile. In group sessions, she has become a frequent helper and supporter of others. She is still not fully back to herself, but she is much improved. Importantly, as Eric continues to benefit from the same process of optimization, he ensures that he will be in a position to continue to help her as the years go by.

I'm inspired by this story because there have been many times in which I have noticed that patients' spouses treat ReCODE sort of like a pill that has been prescribed to their loved ones, rather than a comprehensive protocol that will be far more successful if everyone is working together. Unfortunately, I have even seen cases in which a skeptical partner repeatedly interferes with any attempt at improvement. In much the same way as people are more likely to quit smoking if their partner also quits, more likely to exercise if their partner also exercises, and more likely to make positive dietary changes if their partner does, too,[33] it is clear that when people are supported by a partner they are more likely to succeed on the path to a longer brainspan.

Much of the research on this phenomenon suggests that these successes stem from some combination of social pressure, accountability, and the ease of making a change when it's already happening in one's household. I agree with all of that. But I have also come to believe that there is another force at play: the immediacy of one's "why."

Whenever the going got tough for Eve, she didn't need to look far for her "why." It was Eric. And when Eric struggled to remember why he needed to change his life, his answer was right there by his side. It was Eve.

MY WHY IS SIMPLE

My why is a simple one. You know how, when you see a house that is only partially built but has grand potential, you can't help envisioning what it could be on completion? And thinking how much you'd like to see the finished dwelling? Well, I've seen the brick-by-brick progress over decades as we have seen neurodegenerative diseases move from abstruse to rational, understandable, predictable, treatable, preventable, and ultimately optional. Our data and clinical experience show clearly that these diseases, and brain aging as well, are now ripe for eradication. We need global programs, just as were employed against smallpox and polio, to offer everyone the opportunity for a lifetime of excellent brain function.

And I want to be around to see that. Each of the failures I've witnessed has stuck with me, and I frequently wonder what we could have done better for each of them. I want to be able to tell their children, our children, and everyone that the horrors of neurodegenerative diseases are no longer the concern they have been until now.

WHAT'S YOUR WHY?

In the next chapter, we'll begin setting goals and identifying waypoints for reaching those goals. After that, we'll begin laying out the concrete steps that you will most likely need to take if you want to realize the possibility of having an ageless brain. In other words, the hard work is about to begin.

So, before we go further, I'd like you to take some time to figure out your "why."

What is your purpose on this Earth? What is the cause for which you fight? Why would you be more able to support that principle, that commitment, if you were not worried about or suffering from memory loss?

What is the reason for your life? Why would you be better able to support that impetus for living if you could, in fact, live better and longer, with a mind that is fully prepared to meet the challenges of not just the now but all the years ahead of you?

If you are able to answer these questions, you will very likely have a "why" with which to begin the most challenging parts of this journey. But please do not fear if your answer feels incomplete right now. Keep asking these questions—especially as times get tough and you are tempted to give up. Then, when the demons of denial come to whisper in your ear, tell them why.

And then tell yourself—again and again, if you must. For where there is a why, there is a will. And where there is a will, there is a path to an ageless brain.

6

A MEASURED APPROACH

One only sees what one looks for. One only looks for what one knows.

—GOETHE

Most people intuitively understand that it's often beneficial to have more information.

In academic research, for instance, a comprehensive review of existing literature allows for a better understanding of the subject at hand. In financial planning, detailed knowledge of income, expenses, and financial goals enables the creation of more precise plans for a person's budget and investments. In sports, having more information about an opponent can help a coach devise a more effective game plan.

This helps explain the recent trend toward more wearables—rings, bracelets, watches, on-skin patches, and other devices embedded with built-in sensors to gather and track biometric data. Even though many of these devices have not yet been scientifically validated to extend our healthspans, it makes sense to us that the data they collect can be useful for our wellness, and so they are flying off the shelves.

We can now use these tools to check numerous parameters that warn us of impending disease long before we develop symptoms. We can check heart-rate variability, blood pressure, and even central blood pressure, offering us a meaningful window into our state of stress. We can check body temperature to catch fevers early, warning of incipient COVID or other infections. We can maintain a continuous monitor and record of heart rate and rhythm, oxygen saturation, sleep stages and times, glucose, and ketones.[1] To better monitor aging, we can now easily check the length of protective caps on the ends of our chromosomes known as telomeres, follow the waxing and waning of the multiple microbiomes on and in our bodies, and track the methylation patterns of our epigenome. We can measure the microbes and toxins around us, such as the molds that grow in our homes. We can even obtain a state-of-the-art ultra-sensitive test for the earliest stages of Alzheimer's, delivered right to our homes.

But right now, most of the people who engage in this level of data collection are doing it on their own, at their own expense, without the benefit of knowing what all their data might mean in context. This is because many physicians are still unaware of how the collection and utilization of these data can offer them an unparalleled view of what factors caused, and are driving, a patient's illness. What's more, they don't realize that these data can help them advise patients who are not yet symptomatic of what they can do to stay that way. We really do need to change that, because when it comes to the brain, there is a simple rule: The length of the brainspan is directly proportional to the size of the dataspan.

Thankfully, doctors are increasingly coming to see the benefits of these data. It's certainly not a majority, yet, who will enthusiastically say, "Yes, I'd love to see the information you've been collecting." But, with every passing day, more physicians are coming to understand the true benefits that this biotracking revolution can offer—especially once they start tracking themselves!

For now, though, the truth is that you might have to work to find a doctor who supports this part of your journey toward better brain health right now and sustained healthy cognition for the rest of your life. In some cases, you might be on your own. But while I absolutely want to see a day in which all doctors are enthusiastic about helping their patients understand their data in context, I don't think anyone should assume they're not capable of using their own data to better understand their bodies and brains. And to this end, there are a few "best available variables" that, when well understood and actually acted upon, can make a tremendous difference for our lives, enabling us to better optimize the use of the interventions that we'll discuss throughout the rest of this book for the purpose of improving our cognition right away and keeping our brains fully functional for the rest of our lives.

A BODY OF EVIDENCE

As we have previously discussed, there are six major processes that impact brain aging and neurodegeneration: energetics, inflammation, toxicity, trophic support, neurotransmitters, and stress. So, first things first: We want to get an accurate view of each of the biomarkers that reveals the status of these major contributors.

You might immediately notice that none of these tests is a direct measurement of the brain itself—which is, of course, quite well protected and should be left alone as much as possible. But it's important to remember that, unlike some memorable characters from science fiction—like the villain Krang from the Teenage Mutant Ninja Turtles comics franchise or the character Vagrant from Sergey Snegov's Humans as Gods trilogy—we are not *just* brains! Our brains are an important part of our bodies, but still just part of our bodies, functionally dependent on what happens in all those other parts. So, these measures of what is happening in our "system of systems" as a whole are of great use when it comes to ensuring longer brainspans. The following tests should be part of everyone's annual health assessment, regardless of how healthy one perceives themselves to be.

A Measured Approach

Glucose Metabolism

Glucose metabolism is crucial for brain energetics and can affect all six of the major processes listed above. We've previously discussed the ways in which energetics are so important to the long and healthy operation of our brains, and we've touched upon why the metabolism of the main fuel source, glucose, is so badly dysregulated in so many people. It's thus beneficial to know whether our bodies are suffering from any degree of insulin resistance, and to track the trajectories of this variable over time. Given how many very young people suffer from insulin resistance, primarily as a result of the horrendously unhealthy, sugar-laden foods and foodlike substances we feed our children, and given the ways in which young bodies are so good at hiding the outward effects of creeping metabolism dysregulation, I've long believed that insulin resistance should be measured at annual wellness visits for every child. Furthermore, every adult should be measuring the way in which their body uses insulin.

There are a few relevant tests for doing so: fasting glucose, hemoglobin A1c, and fasting insulin. I will provide ideal brainspan-optimal ranges for these in a chart in chapter 14, but it's important to know not just numbers but what those numbers mean.

For fasting insulin, I prefer to see people on the lower end of the putatively "healthy" reference range. Now, you might reasonably wonder why it would matter. Healthy is health, right? Well, not quite. Reference ranges are typically derived from applying two standard deviations above and below the average scores of relatively small groups of people. And even if those small groups were sufficiently representative samples of the population (and that's almost never the case), it's often an unhealthy population. We are an unhealthy society; therefore our reference ranges are unhealthy ranges!

Hemoglobin A1c, or HbA1c, is a complementary test that measures the average amount of glucose attached to hemoglobin, the protein in red blood cells that carries oxygen. Again, I really prefer to see my patients on the low side of the range commonly considered to be healthy.

For fasting glucose, the lower side of the range is also good, but too low isn't. Why? To answer that question, it's helpful to recall the uphill bicycle analogy we used earlier in this book. In that example, we supposed that we had *just enough* energy to get up the hill, so long as we didn't hit any tire-puncturing burrs along the way. But if we had less energy to begin with, even if we avoided the puncture vine, the result would be the same: We wouldn't make it. Our brains, after all, are energy-ravenous structures and we simply cannot starve these organs of fuel, and glucose *is* that fuel.

If you have a strong family history of type 2 diabetes, you may wish to add an oral glucose tolerance test, with insulin levels, since that is a highly sensitive test, evaluating your insulin response to a glucose challenge instead of simply a fasting level.

We follow a typical stepwise pattern as we develop insulin resistance. First, the glucose tolerance test shows increased insulin response. Next, fasting insulin increases. Then, as your insulin can no longer keep up with demand, your hemoglobin A1c increases. Finally, your fasting glucose increases. What this means is that we can almost always see insulin resistance coming years ahead of time. And if we can do that, then we can reverse it. That means a brain that stays healthier from right now until the end of your very long life!

Because avoiding insulin resistance is so important to long-term brain health, I've been perplexed at the level of resistance I've seen to continuous glucose monitoring (CGM) as it has become more common and accessible. When it first arrived, it was considered to be for diabetics only—and so many doctors still think this way that it can sometimes be difficult to get a physician to prescribe a monitor. To me, this is like saying that chest X-rays are only for those with lung cancer. That's an outdated attitude, focusing on disease (and reimbursement) rather than wellness. Thankfully, many doctors are coming to see the benefits of having people who don't have diabetes wear a CGM (and over-the-counter versions have recently become available, such as Dexcom Stelo), typically for two weeks, so that they can see when their glucose spikes, and when it troughs—both of which are damaging to the brain. While high-carbohydrate diets cause this roller-coaster in most people (as we'll further discuss in chapter 7), everyone responds differently to different types of food and patterns of eating, and these responses can significantly change over time, so wearing a CGM for a few weeks every few years can provide great data from which to begin working to smooth out those peaks and valleys.

Ketones

Glucose is only one of two substances that our bodies and brains use for fuel, and when it is not as readily available, we additionally—and sometimes even primarily—derive energy from stored fat in a process that can be monitored by measuring the blood-circulating ketones that are byproducts of this particular kind of metabolism.

We know this fuel source is quite important to cognition, in part because of the pioneering work being done by Prof. Stephen Cunnane, a clinical researcher at the University of Sherbrooke in Canada, who has demonstrated that ketone supplementation alone can improve brain function in people with mild cognitive impairment.[2] It can generally be inferred that these ketones are created, in a process called ketosis, when we have a low baseline HbA1c and insulin, especially when we fast from food (another topic we'll be covering in the next chapter). But the relationship between HbA1c and ketosis isn't always intuitive, especially when a person's ability to metabolize insulin is already dysregulated.[3] That's why it is valuable to measure circulating ketones.

We can all easily measure our ketone levels with a simple finger stick (which measures one ketone: beta-hydroxybutyrate) or a Breathalyzer such as Biosense (which measures another ketone: acetone). These levels will vary throughout the day—during exercise, while fasting, and in response to different kinds of foods, especially carbohydrates. Because of this, it's good to test at various times. That's a lot of finger pricks!

Fortunately, as I was writing these words, research and development teams from around the world were working hard to get a continuous ketone monitor onto the market just like the continuous glucose monitors I mentioned above.[4] I am confident this will be a positive development for people who want to understand how their bodies move back and forth between utilizing glucose and ketones.

When you can do that easily, you are by definition "metabolically flexible," and your brain will again bless you with improved performance and longevity. I like to see when my patients are able to easily shift into low-level ketosis at least once per day—but I don't want to see them in a state of high-level and unremitting ketosis. Our brains need fuel, and a balance of low (but not dysregulated) glucose levels and high (but not starvation-driven) ketone levels is ideal.

Apo B

Because there is a long-established association between vascular disease—diseases of the vessels of the circulatory system in the body—and Alzheimer's disease,[5] dementia,[6] and cognitive decline,[7] it's critical to monitor blood levels of Apolipoprotein B-100, or Apo B, which transports "bad" lipids, like low-density lipoprotein, across the body. Since Apo B appears to be strongly predictive of cardiovascular disease,[8] and cardiovascular health is predictive of cognitive impairment,[9] this biomarker—high levels of which should be cause for action—may be something of a forward-deployed sentry for brain health at any age.

Bucking many other disconcerting trends—and for reasons that aren't completely understood—Apo B levels among young Americans trended healthier during much of the first two decades of this century. Yet as Northwestern University pediatric cardiologist Amanda Perak and her colleagues noted in a report revealing this promising development, it was estimated that as of 2016, the end of the study period, only about half of US youths had ideal lipid levels.[10] So it is that I still believe that this is one of the measures that is valuable across a person's lifespan, and certainly from early adulthood.

Hs-CRP

We've previously discussed why brain inflammation is so bad for cognition. But of course, inflammation does not *only* happen in the brain. That which occurs anywhere in the body, especially a lot of it over time, can wear down the vascular blood-brain barrier, which is vital for the protection of neurons while signaling the presence of infection elsewhere in the body to stoke a healthy inflammatory response, thus making chronic inflammation a proverbial vicious cycle.[11] This is why it is important to also measure inflammation across the body, and to that end, high-sensitivity C-reactive protein, or hs-CRP, is another test that offers tremendous preventative value, with lower levels indicating very low systemic inflammation and higher levels being a cause for concern and action to stem the tide of an overactive immune system. As multiple studies have suggested, 10 percent or more of young adults may have very high levels of hs-CRP (so high, University of North Carolina psychologist Lilly Shanahan has noted, that they would be dropped from many studies on chronic physical and mental illnesses).[12] This is another measurement that I would advise as part of an annual regimen starting in your twenties.

Homocysteine

The final measurement of the human system-of-systems that I strongly recommend is an annual homocysteine test, which measures molecules produced in the body as a byproduct of the metabolism of methionine, one of the essential amino acids, which is found especially in meat, fish, and dairy products. When homocysteine levels are too high it is an indicator that our bodies are not properly processing these proteins, and although the dynamic consequences of this dysfunction is complex—with resultant effects on vascular cells, cellular oxidation, neurotoxicity, and epigenetic abnormalities—the end results all move in one direction: toward cognitive decline, brain damage, brain atrophy, and degenerative dementias such as Alzheimer's disease.[13] Reducing these high levels has been shown to prevent further damage. Thus, this is another systemic sentry test that I recommend annually.

A FORTUNE TELLER FOR YOUR BRAIN

So far, we've discussed markers that can tell us what's happening inside our bodies right now that, in turn, offer us insights into what's concurrently happening inside our brains. In some cases, such as in the case of glucose, these biomarkers can act as systemic sentries, warning us of imbalances in our bodies that will eventually put our brains at grave risk of diminished functioning.

There is another set of tests that are quite specific to the function of the brain itself—tests that I believe offer tremendous preventative value. These come in two groups. The first group tells you *if* you have any changes of Alzheimer's or advanced brain aging. The second group tells you *why* you are developing Alzheimer's or advanced brain aging. For someone seeking a hundred-year brainspan—one in which no perceptible cognitive decline ever occurs—these are the tests that alert us to what really matters. For that reason, I'm going to spend a little more time describing what they are and how they can help alert us to potential problems in our brains.

GFAP

The cells in our brains form a team: The neurons are the communicators, the astrocytes are the protectors, the oligodendroglia are the cable guys, and the microglia are the cleaners. So, when insults occur, the protectors—the astrocytes—spring into action, and what is remarkable is that you can measure that in the blood! Just as hardworking people produce sweat, hardworking astrocytes produce GFAP (glial fibrillary acidic protein), and this then enters the bloodstream. Thus, we can measure GFAP as an indicator of ongoing brain inflammation and attempted repair.

Blood GFAP measurements have the potential to contribute to accelerated diagnoses of many of the diseases most associated with dementia.[14] The widespread appreciation of GFAP as a biomarker of tremendous value got a big boost in 2023, when it was found that GFAP levels begin to spike ten years before any of the classic symptoms of Alzheimer's disease.[15] At the point in which GFAP is starting to rise, many people may be just barely beginning to recognize the early signs of cognitive decline, and they might not yet even be able to be diagnosed as having mild cognitive impairment. After all, these signs are all too often described as a "normal" part of aging, and such is the power of the whispering demons, that people in this situation may be disinclined to even seek a diagnosis to begin with. But additional research tells us that GFAP measurements at this early stage are a powerful prognostic tool for both the progression and ultimately severity of cognitive deterioration.[16] GFAP has turned out to be quite sensitive, but not specific—since it increases in other diseases, such as the prion disease, Creutzfeldt-Jakob disease, and Alzheimer's—but we're learning more about it all the time.

When I first coined the term *cognoscopy*, it was because I wanted people to associate it with the other test everyone should get at age fifty: the colonoscopy. I have never met a person who *wants* a colonoscopy, but increasingly people understand that it's something they *should* do. (There are new tests that can help many avoid a colonoscopy.) Since far more people will ultimately suffer from neurodegenerative disease than diseases of the colon, I sought to encourage people to be even *more* proactive about assessing brain function. I initially suggested getting a first cognoscopy at age forty-five, but I soon realized that by that age, many people at risk of profound neurodegeneration have already suffered from a litany of cognitive insults and are already experiencing the network of insufficiencies that result. Indeed, as I mentioned earlier in this book, I am seeing younger and younger people with cognitive decline. So it was that forty-five turned into forty, and it wasn't too long before I began to suspect that forty would be too late for many people as well. As more data continue to come in, it seems that, to truly prevent cognitive decline and preserve a very long brainspan, thirty-five is a better starting place for tests like GFAP.

Since I ultimately concluded that forty-five might be too late, then forty might be too late, you might logically wonder whether thirty-five is also too late. Might thirty or twenty-five or twenty be even better ages to take an assessment of GFAP? I have asked this question, too, and the answer is: Earlier might be better for some people. After all, we know that the human body at large experiences a lot of aging during early adulthood, so you can make this choice for yourself. (At the time of this writing, a GFAP test could be obtained for about $150.) For now, I recommend a baseline GFAP at thirty-five and then every five years after that, and you can get this as part of a suite of complementary tests: GFAP, p-tau 217 (see next section), and NfL. Even if the first test shows your levels of GFAP are low, test every five years thereafter, even if you don't feel as though you are experiencing any symptoms of cognitive decline.

If your result is high, please don't worry—you can follow it as it returns to normal as you treat the underlying inflammation with the strategies that we will discuss later in this book. It's likely that, at some point, even if you are taking exceptionally good care of yourself, you will see some rise in GFAP, so it is important to note that elevated GFAP itself is not a death sentence. It does not even mean that you are *likely* to suffer from cognitive decline, let alone neurodegeneration. This is, after all, an indicator of inflammation in the brain, and inflammation at its heart is not a bad thing—it is our body's natural and healthy reaction to the many insults we suffer as a result of living on this planet. When it comes to preserving our chances of achieving a hundred-year brainspan, it is *chronic* and *elevated* inflammation that we are worried about. This is why, upon any significant symptoms or any substantial spike in GFAP, yearly testing (instead of every five years) becomes sensical. Upon *any* diagnosis of any level of cognitive decline, it may be valuable to begin assessing this measure even more often, although my experience is that it generally takes about six months of intervention to lower chronically elevated GFAP, so it makes little sense to test more frequently than annually, at least at the onset of a fight to reverse cognitive decline.

P-tau

I am always fascinated by the elegance involved in brain signaling, and you can see it in all its splendor whenever the brain responds to an insult, instantaneously shifting from "connection mode" to "protection mode," a process that results in a shift in the structure of tau, a protein that plays a vital role in neuronal stability. Not surprisingly, then, when there are many ongoing insults, there is a lot of this changed form of tau, known as phosphorylated tau, lying around! And you can measure this increase in phospho-tau in the brain, in the spinal fluid, and even in the blood.

Now, it might surprise you that I'm interested in measuring tau, because I have been dismayed by the fact that we have spent so much time, money, and hope on the very popular but very broken notion that amyloid plaques and tau tangles are the causes of Alzheimer's disease. We've also failed, again and again, to demonstrate that eliminating these pathologies will result in any meaningful quality-of-life improvements for people suffering from dementia. But they are important biomarkers of Alzheimer's and advanced brain aging. Indeed, I believe phospho-tau is a particularly good substance to monitor for anyone who wishes to achieve a hundred-year brainspan.

There are many different sites on tau that can be phosphorylated—a phosphate-and-oxygen group is attached to the tau protein—but the one specifically associated with Alzheimer's is p-tau 217, so named because the phosphorylation is at amino acid number 217 in the tau molecule. (P-tau 181 is almost as good a marker as p-tau 217 and it is also commonly tested; either of these is sufficient, although I do prefer 217 for its greater accuracy.) The original p-tau 217 tests were not very sensitive, so the increase would not be detected until well after symptoms started, but there is now a highly sensitive p-tau 217 test that shows increases long before symptoms appear (this is from Neurocode, and uses a special ALZpath machine, with SIMOA technology). Even more sensitive ones are on the way and, at the same time, these tests are becoming more accessible, too. In 2024, for instance, a team of experts from ten nations on four continents announced that they were working together to develop a finger-prick p-tau 217 blood test that is thought to be about 97 percent accurate in identifying the presence of disease-associated tau.[17] Several other efforts to create equally accurate and accessible tests were underway at that time, meaning a race to introduce a cheap and easy test for people across the world was inevitable.

When that happens, there will be millions of people who will be prompted to say, "Oh, I was feeling fine, but my p-tau is high, so I guess it's time to do something about it." They'll have an incentive to act. From my perspective, that's not as good as preventing *any* level of disease-associated tau buildup to begin with, but I'm not going to find fault in favor. Right now, after all, most of the people I am consulted about are *already* suffering from some degree, and often a considerable degree, of cognitive decline, and we have seen tremendous success testing that group. The advent of widespread testing for tau and other cognitively associated biomarkers (for instance, an amyloid-beta 42:40 ratio test, which is a measurement of the other substance that is well associated with Alzheimer's disease, can offer similar information to p-tau 217) is going to influence millions of people to take action sooner, which makes the prospect of success in preserving brainspans even greater.

Most people with an elevated p-tau will have earlier experienced a spike in GFAP but that's not always the case, so these tests are complementary and I thus recommend them at about the same rate—an initial test to derive a healthy baseline at age thirty-five, followed by a five-year check until symptoms or other tests prompt more frequent testing. I should probably also accentuate, however, that if you haven't gotten these tests done yet, and you don't happen to be that young, you're certainly not out of luck. As a reminder, the armamentarium for improving tau and preventing dementia is now vast. We have had tremendous success in reversing the symptoms of cognitive decline among people whose first experience with a GFAP, a p-tau 217, or any of the other cognoscopy tests came *long* after the onset of symptoms. So, getting these tests done at *any* age before symptoms accumulate, or, even better, before they begin, is a step in the right direction.

NfL

When Johns Hopkins University neurologist Leah Rubin and her colleagues decided to investigate the impacts of repeated head collision on college football players, she enlisted the help of the NfL.

That lowercase *f* is important, for I am not speaking of the National Football League. Indeed, the NFL, with a capital *F*, has been infamously reluctant to acknowledge the brain damage suffered by former players—and was even initially obstructive of efforts others took to investigate the connection.[18] Rather, the NfL I'm talking about, and which Rubin's team employed for this study, is a neurofilament light chain test. Neurofilaments provide structural support inside neurons, and there are neurofilament heavy chains (NfH), medium chains (NfM), and light chains (NfL), the latter of which are the structures at the heart of this blood test. By using an accelerometer-embedded mouthguard to measure the frequency and magnitude of head impacts on players during a single practice, the researchers found that faster and harder hits were associated with increased NfL levels.[19] In other words, brain damage could almost *immediately* be measured by a sudden increase in NfL.

NfL isn't just a measure of immediate injury, though. It's also been shown to be a robust biomarker for several neurodegenerative conditions, including Alzheimer's disease, MS, frontotemporal dementia (FTD), and ALS. That's why it is another of the tests I include in a cognoscopy, advise other physicians to provide for theirs, and encourage anyone to consider as part of their efforts to catch their own potential neurological problems as close to the wellness-to-disease point of inflection as possible.

Whereas GFAP measures inflammation and repair, and p-tau 217 measures synaptoclastic signaling (pulling back and removing synapses) in response to insults, NfL is an indicator of *actual* neuronal damage, which can occur independent of chronic inflammation and elevated p-tau, as indicated by the acute head traumas being investigated in the study of college football players. It is also, obviously, a test that we would hope we never have to respond to, since GFAP and p-tau should almost always give us opportunities to intervene at earlier states of distress. Nonetheless, while elevated NfL is a very serious matter, it also offers us some opportunity to take action to prevent a little damage from becoming a lot of damage.

NfL also represents chronic brain injury that has accumulated over time as a result of the never-ending march of insults we discussed previously in this book. For this reason, NfL testing might be delayed for a little longer than GFAP and p-tau 217. On the other hand, like those measurements, it may be wise to understand your healthy baseline long before symptoms of disease set in. And since NfL can be deeply affected by physical brain trauma, individuals with a history of head injuries, like those who have played contact sports or suffered concussions, may wish to track it even earlier and more often.

Scans

With those three simple biomarker tests, we now have a remarkable amount of information about the goings-on in our brains. GFAP tells us whether there is ongoing inflammation. NfL tells us whether there is damage to neurons. And p-tau 217 tells us whether a person's brain is responding to insults with Alzheimer's-type signaling. In the vast majority of cases, that information alone—particularly when combined with the other biomarkers I mentioned above—is enough to act on.

Sometimes, however—especially in the case of unexplained symptoms that aren't accompanied by "red flag" levels of these biomarkers—we will need to undergo magnetic resonance imaging (MRI), positron emission tomography (PET), or single-photon emission computed tomography (SPECT) scans, which can offer a glimpse of any actual structural changes in our brains.

An important part of this process is volumetrics, in which a measure of the three-dimensional space taken up by key brain structures—such as the hippocampus and ventricles—can be compared to past images or, in lieu of that personal record, standard norms. These images can also indicate areas of reduced blood flow or problem areas in the brain's myriad connections that can contribute to accurate diagnosis.

A WISH FOR VERY BORING MEASUREMENTS

Recent years have seen an explosion in biomarker tests and, quite importantly, vastly expanded availability of these tests. Whereas at one time nearly everyone was fully reliant on a doctor to order a test—albeit generally only if an insurance company would cover its costs—most of the tests I've recommended here are either already available as measurements that can be taken at home or are soon to be substantially available online or in community labs at prices that most people can afford. In the coming years, I am quite confident, the entire list of tests I've mentioned in this chapter will be available for less than what many people spend each year on coffee.

There is one more test that I think you will find valuable—one of the easiest entry points into measuring the myriad factors that contribute to brain health. It is a test I mentioned earlier in this book, the Montreal Cognitive Assessment, or MoCA. There are several versions of the MoCA available for free online, so there are few barriers to using this test to check in on yourself every now and then. However, I would also advise having this or another cognitive examination professionally facilitated by a psychologist every five years starting in one's mid-thirties, even if you do not perceive any symptoms of cognitive decline—and earlier than that if you do perceive such symptoms.

Not every psychologist is legally able to provide a MoCA, because the test's owners now require payment for a license to administer it. Thankfully, there is a very similar test, the Saint Louis University Mental Status (SLUMS) exam that is available for free and which has been shown to be comparably useful for the detection of cognitive impairment and dementia.[20] An even more sensitive cognitive test is the Cognitive Neuroscience Society Vital Signs (CNSVS) test, which we have used in our clinical trials to complement MoCA and extend the dynamic range of cognitive scores. Other researchers and practitioners include the BrainCheck digital cognitive assessment platform, the assessments developed by Cambridge Cognition, or other similar tests. There are scores of additional and potentially valuable tests for someone who is experiencing symptomatic and significant cognitive impairment. Any of these can be a useful place to *begin*, but none should be where your explorations end, even if they continuously indicate good cognitive health, because these tests measure *symptoms* of aging and disease—and what we want to do is get way ahead of all that! This is also what the other tests I've mentioned in this chapter provide.

We could, of course, do even *more* testing. Indeed, we could all test ourselves right into bankruptcy. But the measurements I've detailed in this chapter will identify the vast majority of cases of cognitive decline, likely a very long time before the demons of denial have a chance to start whispering. As long as we are willing to listen to our brains, and not those demons, the very few cases that slip through this network of sentries will be identified when the earliest symptoms appear. At these points, cognitive decline is not difficult to stop and reverse.

Of course, those reversals cannot happen without effort and action.

A friend of mine who served on an aircraft carrier in the Persian Gulf once told me of his time in the ship's combat information center, where his job was to keep watch over all surface vessels in the region, using radar, reports from reconnaissance aircraft, and other sources of intelligence. This vast network of monitoring tools was only as good as his ability to stay focused on what was actually important, and what was a likely threat. One slipup, he was often warned by his superior officers, could put thousands of lives in danger. Thus, while he had access to a virtually endless trove of information, he kept most of his focus on the vessels that were the greatest danger. Occasionally, when the "all is normal" indicators shifted in one way or another, he would rely on other sources of information to affirm what he was seeing and, on occasion, to advise the ship's leaders on a potential concern. But few people are

A Measured Approach

crazy enough to mess with a US aircraft carrier and the battle group it travels alongside. So, I'm very pleased to report that my friend was bored most of the time.

That's what I wish for you as well—a very boring relationship with these measurements. For that to happen, it will be important to have a willingness to make necessary changes in response to any concerning test results. Even better, though, is to head off most potential problems before they show up on a test. And these two strategies are what the rest of this book is about.

7

EATING FOR AN AGELESS BRAIN

> So long as you have food in your mouth, you have solved all questions for the time being.
>
> —FRANZ KAFKA

When our daughters were young—Tara was five or six and Tess was three or four—our family lived in Southern California. We spent hours pushing them on a green disk swing, where they would fly up nearly vertical, with peals of child giggles. We played soccer against the garage door and watched them ride their trikes and bikes in the driveway, where they would laugh the laugh that is the children's gift to the world.

We had five navel orange trees in the yard and, like everyone else back then, I thought that fresh-squeezed orange juice was very healthy. So, out we would go, past the basketball hoop and into the grove, picking and juicing, and drinking quarts of the stuff.

If I had known then what I know now, I would most certainly have found another activity to share. It takes about four oranges to make one glass of orange juice, and just one orange contains about 12 grams of sugar, so that one glass had nearly 50 grams of sugar in it. About half is sucrose, about a quarter is glucose, and the last quarter is fructose.

These are natural sugars, but when we drink fruit juice devoid of its flesh and fiber, the biochemical effect in our bodies is little different from beverages that have been sweetened with sugar,[1] and there is some evidence that the long-term effects can be just as pernicious.[2]

Even then, I would have cringed to see an adult dump a sparkling white cascade of refined sugar into a child's drink and, given what we now know about the toxic long-term effects of soda consumption, these days I am often aghast when I see children drinking these sorts of beverages. Yet that's not far from what I was doing. And while there remain many unanswered questions about the potential benefits versus detriments of fruit juice consumption over time,[3] on balance it seems it should have been rather obvious to me all along that our species didn't evolve to consume fruit this way, as little more than water and pure sugar.

If we are to protect our brains and our health, we cannot be resistant to the accumulation of new knowledge. And fortunately, we now have quite a bit of information about both the short- and long-term impact of the foods and drinks we consume, and how these affect our health. Of course, this information remains incomplete—we will be learning so much more in the years to come—but the evidence we do have can still lead us to some commonsense conclusions when it comes to protecting our brainspans. So, while we probably shouldn't beat ourselves up over past mistakes made with good intentions, it would be a shame to continue to perpetuate bad habits simply because "we used to do X," or "I have such good memories of Y," or "Z didn't kill me, so it's probably fine."

I'd like to ask you to approach this chapter with an open mind. If you recognize the ways I describe a brain-damaging diet as similar to the ways in which you have personally eaten throughout your life, or how you have fed your own loved ones in a world in which it can be exceptionally hard to avoid all manner of unhealthy foods, please understand that I am not suggesting that you "should have known better." What I wish to offer you is not fodder for lamenting the past but information upon which you can build a path to the future. That is, after all, the only part of our lives we have the capacity to affect for the better. And there is no more powerful way to have that impact than to focus on how the foods we eat affect our brains.

While there are many places where one can start the journey toward an ageless brain, the best place to start is by moving as swiftly as possible toward adopting a brain-sustaining diet. And no doubt you'll see a pattern here. Once again, we want to optimize synaptic biochemistry by addressing the six major players in brain performance and protection: energetics, inflammation, toxicity, trophic support, neurotransmitters, and stress. Nutrition can positively impact all of these—especially if you know what to eat and when.

These days it is often said that there is no "one size fits all" diet that suits everyone's individual needs. At a genetic level, after all, we are all different and thus our diets must be different, too. This is broadly true, but it is also often overstated. There are indeed many ways of eating that can be healthy under *just* the right circumstances. I would be quite hesitant to advise an all-meat diet for most people, for instance, but many people have turned to a carnivorous diet for help with autoimmune conditions, with personal success.[4]

When it comes to optimizing cognition and protecting your brain, however, there is one diet that clearly performs best for most people, most of the time. It is a plant-rich (not plant-only), mildly ketogenic diet, with the following characteristics:

- High in phytonutrients (like polyphenols and anthocyanins)
- High in fiber (both soluble and insoluble)
- High in monounsaturated fats and omega-3 fats from sources like avocados, nuts, and seeds
- No grains or dairy
- No simple carbohydrates
- Wild-caught, low-mercury fish (avoid tuna, swordfish, and shark), pastured chicken, pastured eggs, and grass-fed beef
- Detoxifying cruciferous vegetables, such as brussels sprouts and cabbage

- Large portions of organic leafy greens, avoiding insecticide- and herbicide-laced vegetables
- Some fermented vegetables such as fermented beets or sauerkraut
- Fasting for at least three hours before bed, and at least twelve hours between finishing supper and starting breakfast or brunch

We call this diet KetoFLEX 12/3. It is the foundation for many of the successes we've seen in reversing cognitive decline—and I have also come to believe that it is the best possible starting point for a lifestyle that will lead to a long and healthy brainspan. Here's why:

- More than 80 million Americans—about a quarter of the population—suffer from insulin resistance and have lost use of both of the major substrates of energy creation. Energetically speaking, our brains are sputtering! The reduction in carbs will restore insulin sensitivity, and the reduced insulin allows you to make ketones once again, so you become metabolically flexible.
- The good fats and clean protein smooth out your glycemic curve, so you avoid the peaks and valleys associated with brain aging.
- The soluble and insoluble fiber feeds your gut microbiome and helps to patch your gut lining, improve lipid profile and vascular disease, and improve your glycemic curve.
- The cruciferous vegetables help to increase your glutathione and thus aid in detoxification.
- The many colorful vegetables and fruits provide protective antioxidants.
- The fermented vegetables provide probiotics for a healthy gut microbiome.

- The lack of simple carbohydrates improves your oral microbiome and reduces periodontitis, an important risk factor for cognitive decline.
- Key brain vitamins, minerals, and nutrients such as thiamine (B_1), choline (B_4), pyridoxine (B_6), folate (B_9), cobalamin (B_{12}), vitamin C, vitamin D, vitamin E, and magnesium are all provided.
- The omega-3 fats reduce inflammation and vascular disease, and the DHA supports synapse formation.

Is that too much to expect from a single diet? I don't believe so. On the same day I wrote these words I was in a video meeting with a group of brain health coaches who have been working with physicians to achieve excellent cognitive improvements with many of their clients.

One of the coaches expressed concern about a client who had been declining. That person's HOMA-IR, a measure of insulin resistance, which should be no more than 1.2, was 3.47. His fasting insulin was 13.8, well beyond the target range of 3.0 to 5.5. His fasting glucose was 102, also well above the target of 70 to 90. He had very significant insulin resistance.

Conventionally, most physicians would have considered this individual's vegetarian diet to be quite healthy. But what we saw in the data was a reflection of the fact that he was missing many of the pieces that are vital to youthful brain functioning. He had no metabolic flexibility and could not get into ketosis. As it turned out, he also had untreated sleep apnea, which we have seen is often caused by the inflammation of a high-carb diet. He had biotoxin exposure, which we have found can be mitigated by cruciferous vegetables. So, I could easily see why he was declining instead of improving. I could see what needed to be addressed to turn things around. And I could see that he needed to start with a plant-rich, mildly ketogenic diet to create insulin sensitivity and improve his energetics.

The author Isaac Asimov said, "Life is pleasant. Death is peaceful.

Eating for an Ageless Brain

It's the transition that's troublesome." That's how a lot of people come to feel about the changeover to KetoFLEX 12/3. Eventually you'll feel great, and have better cognitive performance and protection, but it's the transition out of our standard American diet that can be troublesome.

That's because the standard American diet (SAD) evolved to meet the needs of *sales* performance, not *cognitive* performance. The strategists at Big Food have taken advantage of the fact that, as we noted earlier, evolution has selected performance over durability and thus we are programmed to make short-term choices at the expense of long-term survival and cognition. This is why simple carbs are highly addictive—just like cocaine, nicotine, or alcohol[5]—such that, even though we are destroying our brains, hearts, blood vessels, and kidneys, we find it very difficult to stop eating these foods. We are sugarholics! We are ultraprocessed-food-aholics!

But we are not trapped.

It's true that some people will find the addiction so powerful that they need support to break it—indeed, there are even groups like Sugar & Carb Addicts Anonymous, an online fellowship of people who use the 12 steps of Alcoholics Anonymous to gain and maintain abstinence over sugar and undesirable carbs. Most others will find that although the transition from SAD to KetoFLEX 12/3 can be a challenge, they ultimately find themselves enjoying the many flavors and advantages of their new brain-optimizing diet, even before they get the added benefit of increased hippocampal synaptic density (and this greater synaptic density means greater memory capacity).

Struggle is natural. But I do know that virtually anyone can make this change. Again and again, I have seen people adapt to this brain-protective diet, despite the fact that they are already suffering from cognitive decline. Sometimes they also have a psychological symptom known as abulia: the loss of willpower that comes from the inability to carry a thought long enough to identify and follow a purposeful course of action.[6] There are widely divergent levels of clinically identified abulia, but I recognize it as a symptom that may happen to people even in the early stages of cognitive decline—they simply cannot muster as much willpower as they once could. Yet they can make these changes. Ergo, I believe anyone can. Because while it is true that people often feel more motivated to make brain-saving changes once they are confronted with symptoms of cognitive decline, it is also true that it is far easier to make these alterations in our lifestyles before those impairments ever happen. If you do that, you are tremendously less likely to *ever* experience any cognitive decline. In fact, while I would, of course, like you to follow all the principles in this book, I would also concede that those who fastidiously adopt a brain-healthy diet and do nothing else to sustain their brainspan have taken a substantial step toward the goal of an ageless brain.

PLANT YOUR FLAG

I doubt very much that I will be the first person to tell you of the benefits of a diet composed primarily of vegetables, fruits, nuts, seeds, and legumes. You've heard that much before, and you'll hear it again and again, for the truth is that there are few dietary concepts that have been so robustly studied as the connection between plant-rich diets and long-term holistic health, including a decreased risk of heart disease,[7] the prevention of diabetes,[8] and a reduced risk of cancer.[9] So, I understand if you might at first feel as though this section, and perhaps this entire chapter, is "flyover territory," but I implore you not to breeze past this vital principle. I will seek to make it worth your while to spend some time thinking about why eating more plants and fewer animal proteins is particularly and specifically valuable for brain health.

It is important to note here that plant-rich is not synonymous with "vegan" or "vegetarian." For one thing, there is absolutely nothing magical about meatlessness in and of itself. Apple Jacks cereal is vegan. Fritos corn chips are vegan. Hershey's chocolate syrup is vegan. Coca-Cola is vegan. There was a restaurant in Portland, Oregon, that was literally called Vegan Junk Food, where you could order up an animal-free corn dog, a heaping pile of nachos, or a hefty bowl of macaroni in

Eating for an Ageless Brain

cheddar-flavored sauce. I have it on good authority that the food at that place was delicious, but its proprietors rightfully made no pretense that *vegan* necessarily means *healthy*. It doesn't.

Vegan diets *can* be brain-healthy when they are primarily composed of fresh fruits, vegetables, nuts, seeds, and legumes. But so can vegetarian diets, pescatarian diets, and other kinds of omnivorous diets. You may choose to construct your plant-rich diet in any of these ways, provided all your brain-sustaining nutrient needs are being met. The key, though, is truly that you are eating far more plants than animal products, and I would suggest that it is a good goal to be getting at least 80 percent of your calories from plants. (There are some exceptions to this rule of thumb, which I'll delve into later in this chapter, but for now let's stay focused on the multitudinous benefits of eating plants.)

Why do plants have such a tremendous impact on brain health? Well, a big part of the answer to that question—the "low-hanging fruit" in this discussion, so to speak—are the holistic benefits that I mentioned earlier. Our brain is the most essential part of our "system of systems," but it does not exist separately from those other systems. If eating more plants results in better heart health (and it does[10]) then there is a resultant effect on our brains. If eating more plants results in less systemic inflammation (and it does[11]) then there is also going to be a resultant effect on our brains. And on it goes!

But plants also have direct and very profound effects on our brains themselves, in no small part because of the antioxidants that are particularly prevalent in fruits and vegetables. As a reminder, antioxidants are compounds in foods that seek out and destroy free radicals, the unstable molecules that are a natural outcome of cell metabolism but that can damage other molecules, such as the lipids in cellular membranes. While only about 15 percent of the entire human body is composed of lipids, about 50 percent of the brain is made up of these fatty compounds! That's a big part of the reason why, even more than the rest of our bodies, the brain is highly vulnerable to oxidative insult, and this is a key part of the explanation for why antioxidants appear to play a substantial role in protecting, maintaining, and improving brain function.[12] Fortunately, it's straightforward to eat an antioxidant-rich diet—it's baked into just about any plant-rich way of eating. (If you're looking for specific recommendations, though, I'll note that artichokes, beans, berries, broccoli, carrots, greens, and nuts are all antioxidant-rich foods.)

One of the major reasons why plants are such a good source of antioxidants is that many are also a major source of polyphenols—naturally occurring chemical compounds that play a protective role against pathogens and ultraviolet radiation and which are known to neutralize free radicals. But polyphenols are not *only* antioxidants. These small molecules have also been shown to have immune-boosting and anti-inflammatory benefits,[13] likely in part because their size allows them to penetrate the blood-brain barrier, the cellular barricade that protects the central nervous system but also prevents the brain from benefiting from larger potentially beneficial molecules. Thousands of these compounds have been discovered and thousands more are likely still waiting to be discovered but, among the polyphenol groups that have been best studied, the impact on brain health is quite striking. Increased consumption of anthocyanins (found in many blue- and purple-colored fruits and vegetables, including blueberries, cherries, eggplant, plums, pomegranates, dark carrots, and red cabbage), catechins (especially in cocoa and green tea), flavones (in apples, broccoli, kale, onions, and tomatoes), isoflavones (in chickpeas, edamame, peanuts, and soybeans), phenolic acids (in apples, berries, citrus, plums, and mangoes), stilbenoids (in blueberries, grapes, peanuts, mulberries, and raspberries), and curcuminoids (in turmeric and curry powders) has consistently been shown to lead to enhancement of cognitive health, including alertness, attention, focus, learning, and memory.[14]

That might seem like a lot to keep track of, but it doesn't have to be. If you're getting most of your calories from fresh or lightly cooked plants, and those foods are flavorful, and you're eating many different plants each day, you're getting a steady load of polyphenols. There are some other principles that must be respected that will affect what plants are best for brain health (and in which amounts), but truly the best way to begin getting polyphenols into your diet is to follow a slogan that has long been associated with the candy Skittles: "Taste the Rainbow." Of course, candies made of sugar, corn syrup, and myriad chemical additives are a brain drain, not a brain protector, but that shouldn't stop us from appropriating the idea and applying it to the vibrant, healthy plants we eat.

A rainbow of plants might include white radishes, yellow peaches, red raspberries, purple onions, and green avocados one day. On the next day it might be white cauliflower, yellow bell peppers, red strawberries, purple cabbage, and green brussels sprouts. On the day after that, you might choose white mushrooms, yellow lemons, red beets, purple eggplant, and green cucumbers. You get the idea. Not all these individual foods are a heaping source of polyphenols, per se, but this sort of strategy will, over time, make it hard to miss out on the myriad polyphenols that are available.

You might have noticed that there are plenty of fruits mixed into these examples, and I want to continue to be clear about this point: A brain-healthy diet is *not* a no-fruit diet, for as you will recall, resisting cognitive decline is about striking a balance between having the energetics that will get you "up the hill" without an excess that results in metabolic dysregulation, thus adding a flat tire to the journey. I thus remain a big fan of healthy, low-glycemic organic fruits (things like berries, not tropical fruits like pineapples). And while it is true that I rarely find a piece so delicious as that which I used to pick from my backyard trees with my daughters, whenever I can find a truly sweet and juicy orange, I have absolutely no qualms about eating it. But I eat the *whole* fruit, because the fiber provided by the fruit—which is absent in the juice—reduces the glycemic effect markedly.

That's especially true because the sugar that our brains need can be more than sufficiently accounted for by lower glycemic fruits like berries, apricots, peaches, cherries, and avocados. And there are endless vegetables that offer a very low dose of natural sugar but take even longer to break down to glucose in the blood, including leafy greens, beans, lentils, broccoli, brussels sprouts, and squash.

THE PROTEIN QUESTION

The question of whether a purely vegan diet is healthier than one that includes some meat has not been resolved.[15] Even if it was, though, the vast majority of people on this planet eat meat—and for most people, meat remains the easiest and least expensive source for the complete set of essential amino acids humans need to survive. But not all meat is created equal—especially not when it comes to the effect of this food source on our brains. Both quality and quantity matter.

Let's talk about quality first.

One of my early patients had fallen in love with tuna sushi, a meal he ate several times a week for years. When he was diagnosed with early Alzheimer's, he was told that no one knows what causes this disease, and there is nothing that can be done except to take an ineffective drug. He was yet another person written off as a lost cause. But when we measured the levels of toxins in his body (a subject we'll delve into in chapter 11), we saw that the levels of mercury in his body were off the charts. It also turned out that he had mycotoxins, which are biotoxins produced by some molds.

Where did all that mercury come from? Well, the most likely source was his tuna sushi, since tuna is a high-mercury fish. Dementia can rarely be tracked back to one specific insult, but this was certainly a red flag. It was also something we could mitigate. When he changed his diet and his toxins were cleared, his cognition improved markedly.

Not all fish are bad for our brains. Not at all. Large-mouthed, long-lived fish such as tuna, shark, and swordfish concentrate mercury so that our exposure is much greater, but smaller fish such as salmon, mackerel (not king mackerel), anchovies, sardines, and herring (collectively known as SMASH fish) are easy to digest, contain brain-healthy fats that are quickly absorbed into our bodies, are high in many essential vitamins and minerals, and are low in mercury. I also strongly recommend wild-caught fish over those that are farmed and are thus stuck swimming in their own pollutants.

Other great sources of protein include pastured chicken, grass-fed beef, and pastured eggs. In contrast, animals that have been raised in concentrated animal feeding operations sport toxins too numerous to list as well as multiple pathogens. Nonpastured chicken should be avoided due to the likelihood of its containing arsenic and multiple pathogens, such as *Salmonella*. Nongrass-fed beef includes, among other things, inflammatory fats. Nonpastured eggs have sometimes been found to contain flame retardants, of all things, as well as *Salmonella* in some cases.

The quantity question isn't quite so simple, because while low-protein diets have been associated with greater longevity, they have also been connected to increased dementia risk,[16] a reflection of the age-old antagonistic relationship between protection and performance. Protein is clearly important as brain food: The central nervous system can't operate, after all, without amino acids such as tryptophan, tyrosine, histidine, and arginine, which are used for the synthesis of various neurotransmitters and neuromodulators.[17] On the other hand, protein also activates mTOR, an enzyme that plays a central role in governing cell growth in ways that can potentially speed up aging. So, balance is key—and striking that balance is going to be very different for everyone, based on numerous factors.

Recommendations for a balanced supply of daily protein have been as low as 0.8 grams per kilogram (0.36 gram per pound of body weight) to 2.2 grams per kilogram (1 gram per pound of body weight). A recent study in rodents concluded that a diet with 35 percent protein led to the best metabolic parameters,[18] but for humans on a two-thousand-calorie diet, this would mean 175 grams of protein each day, which is much higher than the 50 grams recommended by the USDA.

So, how do you decide how much protein your brain needs? For starters, you should consider your own circumstances. If you are building muscle (including for the treatment of sarcopenia), recovering from surgery or illness, recovering from toxin exposure, fighting inflammatory bowel disease, suffering from frailty, have a body mass index lower than 20,[19] or absorb protein poorly (for example, due to insufficient digestive enzymes or reduced stomach acid, both of which are fairly common in people over sixty and are easily treatable), you will likely need a diet that is higher in protein. This will typically be in the range of 1.2 to 2.2 grams per kilogram (0.55 to 1.0 gram of protein per pound of body weight), which for a 150-pound person would be 82 to 150 grams of protein per day.

For most others, a protein intake of about 1 gram per kilogram (0.45 grams per pound) will prevent sarcopenia and support brain health. For someone of 150 pounds, this would mean about 70 grams of protein: 4 ounces of fish (25 grams), 4 ounces of chicken (35 grams), an egg (6 grams), and a few ounces of nuts or seeds (5 to 10 grams), as one example.

NURTURE YOUR NUTRIENT INTAKE

When you're eating the rainbow, with a little bit of animal protein in support, it becomes very hard to accidentally miss out on most of the essential vitamins and minerals that your brain and body need for holistic health. (If you're eating a vegan diet, it is important to focus additionally on sourcing all nine essential amino acids as well as specific vitamins (such as cobalamin [B_{12}] and D) and nutrients (like choline [B_4]), which are prevalent in meat but need to be sourced piecemeal from plants). Even then, however, it can be helpful to focus specifically on foods that are rich in nutrients that are specifically known to promote the long-term functioning of our brains.

B Vitamins

As an essential starting place, that means ensuring that your diet includes the complex of B vitamins, including thiamine (B_1, which can be found in beans, fish, lentils, peas, and sunflower seeds), riboflavin (B_2, which is prevalent in almonds, eggs, milk, organ meats, spinach, and yogurt), folate (B_9, which you can get from eggs, dark leafy vegetables, peanuts, and liver), and cobalamin (B_{12}, which can be found in clams, liver, trout, salmon, and yogurt). There are a lot of reasons why B vitamins appear to be cognitively protective, but the theorized mechanism that is favored by A. David Smith, the longtime head of Pharmacology at Oxford University who runs the Oxford Project to Investigate Memory and Aging, has to do with the way these vitamins break down homocysteine, a metabolic product of the essential amino acid methionine, which is associated with white matter damage, brain atrophy, neurofibrillary tangles, vascular inflammation, and dementia.[20] Without enough B vitamins, our bodies struggle to dismantle homocysteine. With these vitamins, our bodies easily convert homocysteine into chemicals we need, which is exactly what biologically young bodies and brains do. We also support energy production and mitochondrial function.

Vitamin C
Given the story I told about orange juice at the start of this chapter, you might suspect that I would worry about people getting too much vitamin C. But that was a story about sugar, not nutrients. Indeed, vitamin C is among the best-studied nutrients in association with cognitive function, and the research is very clear. It plays a major role in neuronal differentiation, the process of brain cell development, the protection of nerves, and many other processes that impact brain health and thus cognitive performance. So it is that dozens of studies have demonstrated that higher mean vitamin C concentrations are associated with long-term cognitive function.[21] And while orange juice might not be the best avenue for vitamin C intake, there's little wrong in my mind with an occasional orange that hasn't been separated from its flesh and fiber, nor other high-C foods like broccoli, brussels sprouts, currants, bell peppers, and strawberries.

Vitamin D
It's long been known that vitamin D deficiency is associated with a greater risk of dementia and Alzheimer's disease,[22] which may help explain why

some studies have suggested that dementia mortality may be higher in places that get less sunlight throughout the year, since the sun is the predominant source for this nutrient.[23] Even if you're getting plenty of sun, however, you may nonetheless suffer from vitamin D deficiency if you are a carrier of any of the dozens of gene variants that have been shown to make it harder to absorb and retain this nutrient, and thus may make it more likely that you will suffer from increased rate of cognitive decline as you experience aging.[24] This is one of the many examples of why knowing your genetic self can be a formidable weapon against brain aging (a principle we'll discuss in chapter 14), but even if you haven't engaged in this form of super deep self-discovery, it's easy to "supplement the sun" with eggs, salmon, tuna, and shiitake mushrooms, or D-fortified dairy and plant-milk products.

Omega-3

Omega-3 fatty acids have long been associated with early brain development, which is why pregnant women and women who are breastfeeding are often advised by their doctors to eat extra omega-3-rich foods or take omega-3 supplements. These polyunsaturated fats have also long been known to exert anti-inflammatory and antioxidative effects, which are some of the reasons why they have been shown to improve cognition in older people. For a long time, however, we didn't know much about what happens between these two stages of life. Although it certainly could be inferred, it hadn't been demonstrated in clinical trials. That is beginning to change, however. As researchers like University of Texas Health Science Center neurologist Claudia Satizabal have shown, long before aging makes cognitive decline more likely, healthy middle-aged individuals with higher concentrations of omega-3–derived factors in their blood also had healthier hippocampi—the seahorse-shaped structure in the temporal lobe that plays a major role in learning and memory, and undergoes early degeneration in Alzheimer's. Perhaps owing to that physiological difference, those individuals were more likely to perform well on tests to measure abstract reasoning.[25] This is why it makes sense at every age to double down on foods like wild-caught salmon, mackerel, sardines, flaxseeds, chia seeds, walnuts, and hemp seeds. It's important to note, however, that the long-chain omega-3 fats like docosahexaenoic acid (DHA) and eicosapentaenoic acid (EPA), which support synaptic construction and brain health, are derived from animal sources like fish. The shorter-chain omega-3 fats, like alpha-linolenic acid that are found in nuts and seeds, are inefficiently converted to long chains, so if you do eat meat, the preference should be on getting these healthy fats from animal sources.

Magnesium, Zinc, and Coenzyme Q10
In addition to these vitamins, there are two minerals and one enzymatic catalyst that are worth the attention of anyone who wants to protect their brain against aging—magnesium, zinc, and coenzyme Q10.

Most human adults carry only about 25 grams of magnesium in their bodies—that's about the weight of a standard AA battery—and yet if you completely removed that mineral from our blood, bones, and other tissues, things would go very badly very quickly, especially for our brains. That's because magnesium is essential to hundreds of enzymatic reactions that permit nerves to function, energy to be produced, and calcium and potassium ions to be transported across cell membranes. What this also means is that it doesn't take a lot of magnesium to make a big difference—just a fraction of a gram a day can pay tremendous dividends in brain health. Indeed, when researchers imaged the brains of six thousand healthy middle-aged individuals, they discovered that those whose diets included a half gram of magnesium each day had significantly higher brain volumes—a key indicator of biological aging. Researchers figured that this dietary habit alone equated to a full year of biological brain aging.[26]

There's even less zinc in the human body—about 2 grams, or the equivalent of a small sugar packet—but it plays a key role in the brain, where it has both functional and structural purposes keeping synapses firing and contributing to the ability of our brains to change the activity of those synapses in response to our ever-changing world. Zinc is also crucial for optimal immune responses and for insulin sensitivity. As such, it's not surprising that while deficiencies in zinc are rare in childhood, when this problem does occur in young people it is often associated with cognitive decline and learning struggles.[27] Zinc deficiency becomes much more common as we experience aging. Sure enough, even when we control for many other factors, it is associated with cognitive decline in this stage of life as well.[28] This is why diets rich in oysters, shellfish, legumes, and nuts can be beneficial to lifelong brain health.

We've previously discussed the importance of energetics—the sources of fuel that power our mitochondria. But what would happen if the fuel line were to rupture? Well, that's what would occur if we ran out of coenzyme Q10, or CoQ10, which carries electrons in the mitochondrial respiratory chain. In other words, a boost in CoQ10 can improve mitochondrial function for those who may be somewhat deficient (and one common reason for deficiency is the use of statins). So, it comes as little surprise that researchers have shown that people with higher levels of CoQ10 in their bodies—those perhaps with diets rich in soybeans, broccoli, peanuts, fatty fish, and oranges—had significantly better cognitive functioning and executive function than those with lower levels of this coenzyme.[29]

SUPPLEMENTS SHOULD BE SUPPLEMENTARY

Hippocrates is often said to have remarked: "Let food be thy medicine and medicine be thy food." Alas, the ancient father of medicine does not appear to have ever actually written this,[30] but it remains very good advice. There is a time and place for supplementation and medication, but whenever it is possible to *start* with food, that's absolutely where we should begin.

For some people, though, food alone won't always meet the need, and thus increasing brain-sustaining nutrients through supplementation becomes important. This is best achieved when guided by biomarker assessments, which are increasingly available through private community labs and with at-home test kits. Right now, these tests are usually not covered by private insurance, but that may be changing. Already, several states specifically mandate that insurance must cover biomarker testing for several diseases. As we get closer to a world in which everyone recognizes the clear benefits of catching wellness-to-disease transitions as close to the point of inflection as possible, these tests will be increasingly covered.

The most common brain-sustaining nutrients that some people struggle to get enough of through food alone include those I mentioned earlier in this chapter, as well as menaquinone (vitamin K_2), folate (B_9), iodine, and choline (B_4).

Because supplements aren't generally well regulated, I cannot stress enough how important it is to use suppliers who are committed to purity, safety, and transparency—those who are willing and able to explain exactly where they got their raw botanicals, extracts, and chemicals; how those materials were processed; and how the finished products are tested.

A PROCESS FOR THINKING CRITICALLY ABOUT PROCESSED FOODS

I can't think of highly processed foods without thinking about cigarettes.

For the American Tobacco Company, it wasn't enough to be so big and powerful that any movement of its stock would send shockwaves across the original Dow Jones Industrial Average. The company's top brand, Lucky Strike, may have been the most popular cigarette in the United States, but there was plenty of competition for smokers and soon-to-be smokers. So it was that the company's marketing gurus came upon a new strategy to sell its Luckies: They used images of doctors in their advertisements, infamously claiming that 20,679 physicians agreed that Luckies were "less irritating" than other cigarettes, and that 9,651 doctors believed that "toasted" brands of cigarettes offered "throat protection." Philip Morris International soon followed, claiming its cigarettes were backed by "reports in medical journals by men, high in their profession." Then R. J. Reynolds Tobacco Company got into the game, declaring that "more doctors smoke Camels than any other cigarette."[31]

Most of us have seen these historic ads. Today, they're laughable. It seems difficult to understand how people ever believed that sticking a burning piece of paper, dried up leaves, and chemicals into one's mouth could possibly be healthy. But it's easy to forget just how hard tobacco companies worked to keep this smoke screen going. Cigarette executives weren't fooled by their own advertising. They knew very well their products were causing terrible diseases. Despite this, their companies put chemicals in their products to make them even *more* addictive, guaranteeing that even more people would die from their products. Despite this, tobacco companies infused cigarettes with chemicals to make them even more addictive, virtually guaranteeing that they would harm more people.

I make this observation to point out that while it now seems like utter lunacy that so many people could be so easily convinced that smoking was healthy, even more of us are now being bamboozled by an industry that is just as malicious.

Like the cigarette executives that preceded them, today's purveyors of ultraprocessed foods know that the foods they produce are contributing to—if not outrightly causing—an epidemic in chronic illnesses. In response, they've added ungodly amounts of sugar, salt, and fat as well as other chemical additives that trick our brains to desire these foods even more, leading to soaring rates of diabetes, cancer, dementia, and other diseases.

If it took a major lawsuit to change the practices of American tobacco companies and help lower rates of smoking, might a similar strategy work to mitigate the damages being done to our collective health by ultraprocessed food companies? Psychologist Erik Peper, who teaches and researches at San Francisco State University's Institute for Holistic Health Studies, thinks that approach offers the best possibility to stop an epidemic of chronic diseases caused in large part by the widespread addiction to these foods.[32]

I agree, although I'm not sure our country is ready for that yet—and companies like PepsiCo, Nestlé, and Kraft Heinz have more powerful lobbies than the tobacco companies ever did. What's more, even though we now know about the insidious ways that cigarette companies conspired to keep people smoking, about one in ten Americans still smoke. So, we're unlikely to simply litigate and legislate our way out of this horrendous public health mess we're in. Every individual is going to have to make a choice for themselves about processed foods.

As Michael Pollan, who is best known for his books about the sociocultural impacts of food, often explains it, processed foods are any foods that our ancestors wouldn't recognize as food—which means pretty much everything that comes sealed up in a plastic bottle, aluminum can, or cardboard box. Anything with additives. Anything with food coloring. Anything with stabilizers, deodorizers, or neutralizers. That's all processed food. And your brain is so much better off without any of it.

Whenever I make proclamations like these, I can see people racking their brains for exceptions. What about all those bagged and boxed products in the "health food" aisles of the grocery store? The ones with the brown and green labels stamped with words like "low-sugar," "high-fiber," "plant-based," and "organic"? (Sorry, most of those foods are highly processed.) What about seeds and nuts and dried fruits that come in sealed bags and cans? (OK, maybe you've got me on that one, but check the ingredients. If a bag of almonds includes anything other than almonds, you're already a step away from where your brain wants you to be.)

The solution to all this is to do what the Subway sandwich chain has long asked its customers to do—seemingly without any intended irony. The world's second-largest restaurant chain introduced its "Eat fresh" slogan in 2000, aiming to position itself as a healthier alternative to traditional fast-food restaurants. But the reality was that apart from vegetable toppings, nearly everything else on the restaurant's menu is a processed food, and much of it is ultraprocessed (the difference being that processed foods have undergone minor procedures like chopping or canning, whereas ultraprocessed foods have been fundamentally altered, like chips, sweetened breakfast cereals, chicken nuggets, and artificially flavored crackers).

Really eating fresh means eating vegetables that look like vegetables, fruits that look like fruits, nuts and seeds that look like nuts and seeds, and small portions of wild-caught fish, pastured chicken, pastured eggs, or grass-fed beef.

This isn't simply aesthetic. The link between ultraprocessed foods and brain health has been exceptionally well established, including in a study that followed the eating habits of more than ten thousand people for nearly a decade and found a robust association between eating ultraprocessed foods and poor cognitive outcomes. People whose processed food consumption comprised more than 20 percent of their daily calories had a 28 percent faster rate of cognitive decline than those who ate fewer processed foods.[33]

Why would this be? There are a lot of hypotheses, but a big part of the answer is almost certainly fiber. Processed foods have been cooked up and pulverized down, and thus the fiber—the parts of the food that the body cannot digest—has been compromised. This means nutrients move through the body before they can be used, while carbohydrates get absorbed faster, triggering inflammation and raising insulin levels. Fiber also plays an essential role in the production of short-chain fatty acids—"the messengers from down below," as some researchers have called them—which are blood-brain-barrier-penetrating molecules that serve as a connection between body and brain.[34]

None of this has stopped processed food companies from claiming that their products are "part of a balanced breakfast," or "heart healthy," or some other claim that conveys a false impression of health benefits. Someday, I truly believe, we'll look back on this era of our history with the same sort of incredulousness that we apply to the "doctors agree" genre of old cigarette advertisements. In the meantime, though, it's really up to us to protect our brains against these toxic forces.

GUT CHECK

There's another quality of fiber that is essential to long brainspans: It's prebiotic, meaning it acts as food for the bacteria that live in our guts and which play an increasingly understood and important role in holistic health. That's especially true for our brains, with the diversity of species in our guts having been shown to be significantly and positively associated with cognitive function.[35]

But it actually doesn't matter much if we supply this "microbe fuel" to our guts if there aren't enough microbes down there to consume it. That's why probiotics (as well as prebiotics and postbiotics) are so important to brain health. These are foods like kimchi, sauerkraut, sour pickles, miso soup, kombucha, and yogurt that contain live microorganisms, thus helping to maintain the so-called good bacteria in our guts. That's important because the gut produces many of the same neurotransmitters as the brain—such that it's sometimes called the second brain.[36]

Indeed, in one review of studies on probiotics and cognition, twenty-one of twenty-five experiments on adults resulted in the identification of positive associations over periods from three weeks to six months, and these connections appear to be particularly powerful when it comes to emotion.[37] "While not a direct effect on cognitive performance itself, the limited research currently available indicates that probiotics may provide a buffering effect against stress, meaning that cognitive performance is maintained where it would otherwise be negatively affected," the team, led by nutrition-cognition researcher Jessica Eastwood, wrote.

How exactly probiotics impact emotion—and thus cognitive performance—isn't yet clear, but it is known that bacteria in the gut can produce dopamine, serotonin, and norepinephrine, all of which are known to have profound effects on human mood. What is clear, in my experience with individuals from around the world, is that when we increase the consumption of probiotics, many of the emotions that come along with cognitive decline—fear, anxiety, anger, frustration—begin to mellow, allowing these patients to dedicate even more energy toward taking steps that will lead to improved brain function.

Laurel, a mechanic with cognitive decline who responded to our protocol, is a good example of this. Like many people, the cognitive challenges he was facing prompted some depressive symptoms,[38] which continued even as many of the challenges he was facing were clearly abating. "When I'd have a setback, like if I'd forget something, even if it was really minor, I'd get really, really down on myself," he told me. "I'd get this thought in my mind like 'it doesn't matter anyway and I'm just slowing down the inevitable.' So, then I'd make bad decisions, which only slowed down my progress more. It was really dark."

Over time—perhaps in part because of his improved probiotic diet, although likely as a benefit of that and many of the other changes he was making in his life—these episodes subsided. As they did, his cognitive progress escalated.

"These days," he recently told me, "the only thing that gets me down is when one of my sports teams loses a big game."

THE SUGAR ASTEROID

Back in 1967, most medical journals didn't require researchers to disclose the sources of their funding. So, for decades, nobody knew that a trade group called the Sugar Research Foundation had paid a group of scientists from Harvard University to publish a research review suggesting that the biggest threat to human health was not sugar but fat.[39] It wasn't until 2016 that a historical analysis of internal industry documents by a team of researchers from the Philip R. Lee Institute for Health Policy Studies in San Francisco revealed the payments—and drew a direct line to the results.[40] Those researchers have suggested that the trust that people placed in Harvard scientists and the "all clear" sign the scientists were waving on sugar were a public health asteroid strike, of sorts—a sudden and catastrophic event that led to long-term devastation. Indeed, over the past sixty years, US healthcare costs have gone up about a hundredfold, in large part due to the costs associated with chronic illnesses stemming from metabolic dysregulation, one of the major components of the network of insufficiencies that leads to neurological degeneration.

How many millions of dollars did the Harvard scientists receive for a paper that seems to have contributed to billions of dollars of disease costs, and millions of deaths? Not even one. They were paid $6,500, about $60,000 in today's currency. That is the kind of return on investment that would make even Warren Buffett envious!

In the intervening decades, the sugar industry used their playbook again and again. In 2015, for instance, reporters from the *New York Times* revealed that Coca-Cola had given millions of dollars to scientists willing to promote the idea that most health problems are due not to what people consume but how much they exercise.[41] The following year, the Associated Press exposed the shenanigans of a group of researchers who knew their work was "thin and clearly padded" but who nonetheless went ahead to publish results suggesting that children who eat candy are in some ways healthier than those who do not. And who paid for that work? A group called the National Confectioners Association![42] Others have challenged these sorts of conspiratorial narratives, arguing that scientific research, although funded by peddlers of sugar, has implicated pretty much everything but sugar in the global explosion of chronic diseases, and that the results "are not necessarily the products of malevolence."[43] To me, that's a little like arguing that the dinosaurs actually died by starvation, not an asteroid—ignoring the chain of events that began when a asteroid crashed into our planet with such force that it shook the entire planet and left a crater 120 miles wide and 12 miles deep.

Alas, we cannot undo that cataclysmic event, but, unlike the dinosaurs, we have a choice about what to do next. And, for anyone who aspires to an ageless brain, the path forward is well illuminated: Refined sugar, added sugar (especially high-fructose corn syrup), and natural sugars removed in any way from the plants in which they were sourced must be excised from our diets with extreme prejudice.

MOVING TOWARD MILD KETOGENESIS

Perhaps the greatest tool for supplying our brains with needed fuel while avoiding the overloads of sugar that are ruinous to healthy brainspans is the glycemic index, an estimate of how much specific foods will increase a person's blood sugar levels. Plant-based foods with a glycemic index lower than 35 are best aligned with the goal of preserving and protecting our brainspans.

If you're familiar with using a glycemic index and glycemic load, you may already know that adhering to a cutoff score of 35 would render a lot of common foods as imprudent choices. Bread, pasta, potatoes, rice, and corn are all inadvisable under this standard. This is not to say that you cannot *ever* have these foods (although the grain-based ones like bread and pasta have the additional problem of gut lining damage, so you should indeed avoid these), only that you should be aware of the glycemic load and its impact on your brainspan, and that your decision to accept the spike in blood sugar that will come with that food should be aligned to your "why," as we discussed in chapter 5.

You'll also recall from chapter 6 that glucose isn't the only substance that our brains use for fuel. We also generate energy from the ketones derived from fat (stored fat or fat in the diet) and should be doing so on a daily basis. This is why it is important to measure ketones.

It usually takes a few weeks to get into endogenous ketosis, a state in which you are manufacturing your own ketones from your fat. It can be even more difficult to get there if you are underweight to start. In many cases, I have thus recommended supplementing with exogenous ketones for the first two months or until a person can get their ketone levels to at least 1.0 moles per liter of beta-hydroxybutyrate. These exogenous ketones can be taken as medium-chain triglyceride (MCT) oil or coconut oil, but if you have any vascular disease or risk for vascular disease, it makes more sense to avoid MCT and coconut oil, and instead use ketone salts, esters, or a combination.

Ketosis can also be achieved through fasting. It's likely no coincidence that fasting is one of the most popular foci of researchers studying biological aging. One of the pioneers of this area of research, my colleague (and coauthor of many years ago) Prof. Valter Longo, demonstrated in 2024 that a diet that used specific fasting techniques lowered insulin resistance, liver fat, inflammation, and other markers associated with aging, so much so that Longo and his collaborators estimated that the study's participants had lowered their aging-related biomarkers by two and a half years on average.[44] This was hardly an outlying result; one review of thirty studies of various forms of fasting demonstrated promise in mitigating aging-related challenges related to blood lipids, glucose metabolism, insulin sensitivity, oxidative stress, and inflammation.[45]

It remains difficult to study the human brain in vivo—it's a pretty important and hard-to-access organ, after all (although that is changing with the introduction of new blood biomarkers such as p-tau 217, GFAP, and epigenetics). As such, the research is limited when it comes to whether the brain itself also ages more slowly as a result of fasting. There are, however, many clues that suggest this might indeed be what is happening. Clinical studies have shown that fasting has beneficial effects on the symptoms and progress of epilepsy, Alzheimer's disease, and multiple sclerosis, and animal studies have shown mechanisms by which Parkinson's disease and mood and anxiety disorders may be beneficially affected by periods of reduced food intake.[46]

But even if our brains are aging slower—for the same reasons that the rest of our bodies appear to age more slowly as a result of fasting—that wouldn't be the only factor at play in the ways in which fasting helps promote long-term brain function. That's because, as you may recall, one of the many insults that leads to cognitive decline is a reduction in the availability of energetics that comes largely from metabolic dysregulation. Fasting offers our brains another option to access energy—via fat stores rather than sugar. So, just as I don't believe that a merciless eradication of carbohydrates is necessary or wise, I am not in favor of eating habits aimed at invoking *extreme* ketosis, like the sorts of many-days-long fasts that seem to have become popular among some social media influencers and celebrities.

Remember that reduced energetics is one of the very most important drivers for both brain aging and Alzheimer's disease (as well as Parkinson's disease and others). So, we need more energy, not less—thus fasting would seem to be the worst thing you could do. But the brain energetic insufficiency in brain aging and Alzheimer's is actually due, paradoxically, to excess. That is, excess rapidly metabolized simple carbohydrates, effecting insulin resistance, inability to produce ketones, and thus an inability to use either of the two substrates, glucose and ketones. The brain is literally sputtering like a hybrid automobile running out of both fuel and electric charge. So, we must be careful—we need to reestablish insulin sensitivity (and fasting helps to do that) while avoiding energetic reduction, and that's why metabolic flexibility, in which both glucose and ketones can be utilized, is such an important goal.

The sweet spot for many, and what I believe to be the best starting place for most, is limiting one's window of eating to ten to twelve hours for those who are ApoE4-negative, and eight to ten hours for those who are ApoE4-positive (and, for those who do not know their ApoE status, ten hours). For most people who sleep seven to eight hours a night, that means eschewing food for a few hours after waking in the morning, and declining food for several hours (at least three) before going to sleep at night. For example, if you normally head to bed at 10 p.m., your window for food consumption would begin at 7 a.m. and end at 7 p.m. Those fasting hours, plus a low-glycemic load during the hours in which one is eating, are generally enough to put most people into a mild state of ketosis, although personal experimentation and measurement will help confirm this, at which point adjustments may be made, often in the form of adding another hour or two, at the most, to the fasting window.

CONTROLLING CRAVINGS WITH DRUGS

It might not seem very hard to avoid high-glycemic foods or to go without eating for just a few hours before and after sleeping. Yet these are often the principles of a brain-healthy lifestyle that patients struggle with the most.

That's not because they lack willpower. It's because we evolved to prioritize eating food whenever it was available, whether we really needed it or not, because the timing of a next meal was never guaranteed. This evolutionary programming is so strong that about 20 percent of adults are literally addicted to food in the exact same way that people can be addicted to alcohol and drugs.[47]

In the past few years, there has been an explosion in the availability of drugs that help fight the cravings that make reaching ketosis so difficult for so many people. Semaglutides like Ozempic and Rybelsus for diabetes and Wegovy for weight loss seem like a miracle for many people. As of 2024, about one in eight adults in the United States had been prescribed one of these drugs.

As these numbers have grown, clinicians and researchers have also begun to see some evidence of improvements in cardiovascular disease and prevention of kidney failure, Alzheimer's, Parkinson's, and depression. It will take a lot of time and careful study to better understand why this might be, but the implications are profound: When one targeted treatment affects so many different maladies, it means that these maladies are being caused repeatedly by the very pathway that is being targeted by the drug. By analogy, when it turned out that antibiotics could cure pneumonia, urinary tract infections, meningitis, and other infections, the germ theory of disease was strongly supported. Now, the developing data suggesting that GLP-1 agonists—the class of drugs that includes semaglutides—improve many different conditions supports the metabolic theory of disease and reminds us that our food and lifestyle are truly killing us by these many routes.

If you live in a toxic world, then a detoxing drug will be helpful for many different conditions. Likewise, if you live in a metabolically deranged world, then GLP-1 agonists will appear to be a panacea. And that's exactly what is happening.

These drugs aren't right for everyone. Since they are relatively new, the long-term side effects are unknown and will continue to be unclear for a long time to come. But a recent study suggests that many people won't have to stay on these medications forever in order to reap the benefits. That study suggested that people who stop using GLP-1 agonists but who follow a ketogenic diet are able to maintain their lower body weights and blood sugar levels. In fact, the people who got off GLP-1 drugs but maintained a ketogenic diet *actually lost more weight* than people who simply continued taking the medications.[48]

WE CAN DO HARD THINGS (ESPECIALLY WHEN THEY BECOME EASY!)

I know that I have already reflected several times in this chapter on the difficulty of changing one's dietary habits. It might seem easy enough to move away from meat-heavy meals, focus on brain-sustaining nutrients, renounce processed foods, add plenty of probiotics, prebiotics, and postbiotics, and fast to ensure a mild daily state of ketosis. And the truth is that most people will have very little trouble implementing any *one* of these principles, particularly if they're already motivated by a brush with cognitive decline. It is much more of a challenge, though, to implement *all* these principles—indeed, to dedicate the rest of one's life to them—when the consequences of not doing so do not feel immediate.

With this in mind, I want you to know that it's OK to struggle. It's OK to fail. It's OK to need to try at this, again and again, until all these principles become part of your life. And if you're a parent, it's OK to feel like you're up against the entire world as you help your children develop lifelong good eating habits in a society where sugar and processed foods are so ubiquitous. The most important thing is to keep trying, keep tweaking, since even small changes will help to optimize your brainspan.

Since you have come this far in this book, it is because you are attracted to the idea of a life in which you can emotionally, intellectually, skillfully, and joyfully engage with your neighbors, colleagues, friends, and family members. You want this right now. And you want to keep it going when you are eighty, ninety, and one hundred years old, and perhaps long after that as well. You know this is possible. You know that it is largely in your control. You can take it at your own pace, one step at a time.

There are other ways to help you get there, but none so powerful as what you eat and the ways in which you eat. These principles are "the first bite."

And yes, this can be hard, but there are many workarounds and lots of support from the many who are doing so successfully. I have seen so many people do it. And I know you can, too—because it is worth it.

8

THE BRAINSPAN WORKOUT

> You don't stop exercising because you get old, you get old because you stop exercising.
>
> —DR. KENNETH COOPER

This chapter is just for people in the age range of 18 to 122. If you're younger than 18, please come back when you turn 18, because you are currently too young for the data we'll discuss here. If you are older than 122, you don't need to learn from me, I want to learn from you because you are doing a lot of things right.

Why is 18 our starting point? Well, a recent, quite surprising study showed that a single Alzheimer's risk gene variant, present in 25 percent of the population—the ApoE4 described earlier—affects the memory performance of people *even in the eighteen- to twenty-five-year-old range*. Hard to believe, but true. But please don't worry, because here's the good news: Previous studies have shown that you can obviate the memory impact of this gene variant with one powerful antidote. And if you look at all the published evidence for enhancing cognition (whether you are Gen Z, Millennial, Gen X, or anyone else) and preventing decline, the singular intervention that has the most evidence backing its efficacy is not some billion-dollar drug or expensive procedure.

It's exercise.

To understand why exercise is so good for our brains, it's important to first recognize that before we exercise, we have to have a healthy brain.

By way of example, imagine that you are in a workout session with a personal trainer and she gives you an instruction to shift from one exercise to another. Most of what happens in that moment takes place in your brain. First, sound waves from the trainer's voice cause vibrations in your eardrum, which results in the movement of chemicals at the base of a structure called the stereocilia, prompting the creation of an electrical signal, which is transmitted to the auditory cortex in the brain's temporal lobe. Then, in just a few thousandths of a second, these signals are sent to the parietal lobe of the cerebral cortex to be processed as instructions, and then sent onward to the motor strip in the frontal lobe, to be converted into movement-specific signals that will be transmitted through the spinal cord, to the motor neurons, and thus to the muscles. This remarkable journey, which we so often take for granted, may be interrupted by Alzheimer's disease, which strikes the parietal lobe and often leads to apraxia. What this means is that, although you have the musculature and coordination to carry out an act, you lack the learned set of instructions—you can't link the idea with the movements. This may mean, for example, that you can no longer follow a trainer's instructions.

Of course, when everything *is* working right, this perfectly orchestrated symphony doesn't just happen in the gym with a trainer. It happens when you are running on uneven ground or in a crowd of other runners—transferring quickly changing visual and auditory clues from your eyes and ears to your brain for integration with the dynamic world around you. It happens when you are taking notes in a lecture—rapidly integrating auditory cues with highly developed learned skills like writing with a pen on paper or typing on a keyboard. And it happens when you are riding a bicycle, gathering and rapidly processing endless signals about the world around you to make nearly unconscious decisions about the exceedingly delicate motions in your feet, ankles, legs, hips, abdomen, arms, hands, neck, and eyes that are required to stay upright and moving, often at speeds that the human body was not even evolutionarily designed to maintain.

It never stops happening, either, because our bodies never stop moving. Even when we are sleeping, we are constantly in motion. Our hearts are pumping, our blood is circulating, our eyes are moving rapidly under our eyelids, and our bodies are in a nearly constant state of repositioning, with dozens of major adjustments and hundreds of smaller movements each hour.[1] Every single one of these movements is facilitated by our brains. And yet when people think about brain aging, they almost always think first about functions like memory, fatigue, problem-solving, and comprehension. Movement is often completely forgotten.

What's more, when people think about the reduction of movement that comes with aging, they often think in terms of the atrophy of their muscles and degeneration of their joints. Biological aging most certainly has an effect on these parts of our body, but what is not as commonly recognized is the fact that even in a hypothetical circumstance in which these parts of one's fitness were unscathed, the ability to turn perception into response would still be degraded by aging in the brain.

Thus, it is important to think of cognition not simply as something that happens inside our heads. When we protect our brains, we protect our ability to engage physically in the world—to perceive and respond with movement. And, as it turns out, the best way to protect these vital capacities is to keep using them!

That's why, in this chapter, we will discuss exercise in terms of mechanisms and outcomes—actions that can be taken to engage, strengthen, and protect your motor neuronal signals from aging and the symbiotic benefits for our brains and bodies that happen as a result of those actions. What you'll learn, also, is that there is no singular path to getting all these forms of exercise. You might choose to derive these benefits primarily via weight training and treadmill running. You might choose to get there through soccer and yoga. Given the explosive growth of pickleball as the latest national pastime, I suppose it should be explicitly said that this sport is another avenue through which to strengthen our brains against the ravages of aging.

But yes, you must do *something*. Copious research has demonstrated that exercise reduces the risk of just about every chronic disease—and the adverse outcomes for contagious diseases, too.[2] We now know that the same is true for our brains, and it's very clear that the path to a hundred-year brainspan absolutely must include exercise.

AEROBICS

In the wake of World War II, a US Air Force exercise physiologist named Kenneth Cooper and a physical therapist named Pauline Potts noticed something disconcerting among the young men they were charged with helping train. Military members who seemed to be fit in other ways—those who were trim and muscular, without much fat on their bodies—were often unable to engage in strenuous physical movement for longer than a few minutes.

This, they suspected, was because military training at that time emphasized short-duration drills over long and arduous ones. Together, Cooper and Potts conceived of a new kind of physical training, dubbed aerobics, which was intended to condition the body to better metabolize oxygen for the purpose of creating energy to fuel movement over longer periods of exertion.

Today almost everyone recognizes that aerobic exercise is an essential part of holistic fitness, but many people don't understand its importance for preserving and protecting brainspans—an outcome that starts with the significant role of aerobic exercise in enhancing the oxygenation of the brain.

We know, for instance, that aerobic exercise leads to an increase in cerebral blood flow, the rate at which blood is perfused into brain tissue, which at a very basic level means more oxygen. That's important, because while the human brain comprises only about 2 percent of a person's total body weight, it consumes more than 20 percent of the oxygen we take in, in no small part because while neurons are great at using oxygen, they're not so great at storing it, thus these cells need a continuous supply of the stuff.[3] All of this explains why there is a very close connection between neural activity and cerebral blood flow.[4] What recent research has demonstrated, however, is that this connection begins to falter with aging. Simply put, the aging brain ceases using oxygen as efficiently as it once did.[5]

The aerobic solution to this problem is twofold: helping the brain maintain the efficient use of oxygen and increasing oxygen if and when this efficiency subsides. To the former goal, regular aerobic exercise has been shown to improve overall vascular health, and healthier blood vessels are simply more efficient at transporting oxygen and nutrients to the brain. To the latter goal, aerobic exercise increases blood flow even when vascular health wanes, thus ensuring delivery of the oxygen that brains need to do what brains do![6]

Oxygen isn't just fuel for neural activity. It's also an essential building block for the formation of new neurons, or neurogenesis, and while neurons are amazing cellular survivors—some will stay with us from birth until death—we absolutely can't get by without any new ones at all. This is especially true in the hippocampus, which is crucial for memory, and one of the most important regions of the brain for neurogenesis (and thus for neuroplasticity—the brain's ability to reorganize itself by forming new connections over time, a process that often requires brand-new neurons to serve as a bridge between two other parts of the brain). It thus stands to reason that better oxygen flow will result in improved neurogenesis, which in turn will better sustain memory and plasticity, which also explains why hyperbaric oxygen (supplemental oxygen administered at levels higher than atmospheric pressure) has been shown in many studies to promote new neuron creation[7] and why aerobic exercise has a similar effect.[8] Indeed, hyperbaric oxygen therapy (HBOT) has been helpful for many people, especially those with vascular disease or traumatic brain injury (TBI), and particularly when combined with brief bursts of normoxia, a return to the 21 percent oxygen of the air we breathe. Of course, for all the therapeutic options described in this book, the goal is to start with the most physiological, the most causal—meaning the action addresses the risk factors directly, such as reduced blood flow and oxygen saturation—and the most upstream. Whenever possible, after all, we want to directly address the risk rather than trying to trick nature. And this is what aerobics does.

As we first discussed in chapters 6 and 7, your brain needs either glucose or ketones for energy—and preferably alternates between these fuel sources effortlessly. When glucose is utilized in the absence of oxygen, you obtain a small amount of energy, but it's over tenfold more when you include oxygen, so there is a huge advantage to having plenty of oxygen available. As we also discussed in chapter 7, a low level of ketosis is achievable for most people through dietary choices and eating patterns such as fasting, but we can more easily get to—and maintain—that state of energy use when we combine these eating habits and aerobic exercise,[9] chiefly because this type of exercise helps us burn through the immediate glucose sources of energy, thus sending our systems on alert for the other available sources, namely fat. With that, our brains are benefiting from an infusion of multiple fuel sources instead of just one. In my experience, patients frequently notice that they are able to get into a low level of ketosis (for example, 0.5 mole per liter of beta-hydroxybutyrate) through their diets, but adding aerobic exercise boosts them into that optimal range of 1.0 moles or more per liter of beta-hydroxybutyrate.

While aerobic exercise increases the flow of oxygen into our brains, the other side of this equation is important, too—what comes flowing out? The answer is waste products and toxins, a process that is facilitated through the glymphatic clearance pathway, a recently discovered waste disposal system that helps move soluble proteins and metabolites out of the central nervous system. The clearance pathway was discovered in the mid-2010s by researchers in the lab of the Danish neuroscientist Maiken Nedergaard, and her teams have since demonstrated that while the glymphatic system functions most effectively during sleep, it can likely be pushed into action when we are awake if we engage in regular physical activity, particularly aerobic exercise.[10]

Hiking, jogging, running, and swimming are great ways to increase the flow of oxygen across your body and brain. Cycling and rope jumping, dancing and stair climbing, rowing and cross-country skiing are all excellent forms of aerobic exercise as well. The key is not so much what you do as what that activity does to you. To know that, however, it's important to understand your upper limits of oxygen utilization, commonly known as VO_2 max. Unfortunately, the most accurate way to know your VO_2 max is to take a test in which you are exercising at maximum capacity while wearing a mask that measures your oxygen consumption. The price point for such tests in most places is several hundred dollars, and depending on your means and interest in measuring aerobic fitness, that might be a worthwhile investment for you every now and then. Fortunately, this is a measurement that is reflective of something that just about everyone can intuitively feel and easily track using proxy measurements. If you run a mile every day, for instance, you know how you feel at the end of that mile. And if you run your fastest possible mile, once a week, and track your time as well, you know what your maximum capacity is. If you're interested in putting a number to it, there are various tools and trackers that can help you calculate it within a reasonable proximity, based on your resting heartbeat and maximum heartbeat, including relatively inexpensive smartwatches and biotracking rings.

There are a lot of different ideas about how often you should be engaging in aerobic exercise, but our brains *never* stop needing oxygen, which is why sleep apnea and other causes of reduced nocturnal oxygen saturation remain one of the most common—and most commonly missed by practitioners—contributors to cognitive compromise. I thus believe it is good practice to get some form of exercise every day that

The Brainspan Workout

raises your heart rate and breathing rate (preferably outdoors, to minimize mycotoxin exposure, unless there is significant air pollution—both subjects we'll discuss further in chapter 11). You should consult your physician about making a plan that is right for you, especially if you have underlying health issues, but for most people, at any age, a good goal is to reach one's maximum heart rate, by engaging in up to ten minutes of maximally rigorous aerobic exercise, once per week. Six other days of the week, seek to bring your heart rate up to about 70 percent of your max, sustained for thirty minutes. (Max is typically estimated to be 220 minus your age. So, if you're thirty-five, your maximum heart rate should be about 185 beats per minute, and your 70 percent target would be 130 beats per minute.)

So, the aerobic goal is at least three hours of aerobic exercise each week, and a healthy infusion of extra oxygen to the brain as a result. Three hours might not sound like a lot, but if you're not used to that kind of commitment, it can be hard to figure out how to fit it into your schedule—especially once you recognize that aerobic exercise is just one of five types of exercises that everyone should be engaged in to keep brain aging at bay.

Physician Deborah A. Cohen understands that many people feel that they don't have time for exercise. But the public health researcher, who is also the author of the book *A Big Fat Crisis*, about the growing global obesity crisis, also has a well-earned reputation for calling it as she sees it. And the way she sees it, a lack of time is often an excuse that has little merit.

"There is a general perception among the public and even public health professionals that a lack of leisure time is a major reason that Americans do not get enough physical activity, but we found no evidence for those beliefs," Cohen told the *Washington Post* in 2019,[11] following the publication of a comprehensive analysis of the ways most American adults actually use their time. In reality, Cohen and her collaborators reported, the average American has several hours each day that are unaccounted for by work and family obligations—and we turn over the vast majority of that free time to mobile, computer, and television screens.[12]

While I'd much rather see people fully focused on exercise when they are exercising, I'll also offer this observation: You *can* have it both ways. You can watch a television program while jumping rope. You can scroll through your social media feeds while riding an exercise bike. You can watch your favorite sports team running on the field while you are running on a treadmill. And if you're like the growing number of people who work online or hybrid schedules, you can participate in a Zoom meeting from a height-adjustable desk with a treadmill or walking pad. And if that's what it takes to get you moving toward integrating aerobic exercise into your daily life, I'm all for it—because that's not all you're going to need to do.

STRENGTH TRAINING

It will be many years before we know the full extent of the findings from an intrepid research project called the "Maintain Your Brain" trial, which is testing to see if an individually tailored program of lifestyle modification can prevent cognitive loss in a group of more than six thousand Australians. Scientists involved in such longitudinal inquiries are often and understandably a bit reluctant to "jump the gun" on what interventions work. But when Michael J. Valenzuela, who is one of the leaders of the project, saw what a strength-training program did for a relatively small cohort of a hundred people who had experienced symptoms of mild cognitive impairment, he couldn't hide his enthusiasm.[13] "The message is clear," he said in a statement released by the University of Sydney in 2020. "Resistance exercise needs to become a standard part of dementia risk-reduction strategies."

Needs. That's the word he used. Not "should be considered" or "might be effective." This intervention, Valenzuela said, was an absolute must.

In the study, he and his team assigned one test group of participants to an hour and a half of weight training each week for six months, then continued to monitor that group, along with other test and control groups, for another year thereafter, including conducting periodic cognitive performance tests and three MRIs to measure changes to several subregions of the hippocampus over time. And not only did the weight-lifting participants slow down brain shrinkage and perform better on the cognitive tests, but those effects also appeared to hold on for up to twelve months after the training ended.

I wasn't surprised by this finding—other studies had previously shown that resistance training has positive and often quite powerful effects on cognitive function,[14] which is why strength training has long been one of the ReCODE Protocol's interventions. I was nonetheless very satisfied to see another element of the protocol validated by yet another group of independent researchers, including someone who, like me, seemed to want to scream from the rooftops that we can slow neurodegeneration and often reverse the symptoms.

Why? Well, it should first be said that strength training and aerobics are quite complementary types of exercise. Depending on how you train, weights can contribute to an aerobic workout. When that happens, all the positive benefits to the brain—the oxygenation, the neuroplasticity, the glymphatic clearance, and more—happen at once.

But it has also been shown that there is a strong relationship between strength training and insulin resistance, especially in men—even when those men look very different in terms of weight, waist circumference, body fat percentage, and other demographic and lifestyle factors.[15] This has many holistic health implications, but two are especially important in the context of brain aging. First, bringing insulin sensitivity back toward a natural baseline is a tried-and-true way to reduce inflammation, a major insult contributing to cognitive decline. Second, in most cases, individuals with well-regulated insulin sensitivity find it much easier to reach and remain in a mild state of ketosis, thus they are feeding their brains with a greater diversity of energetics. And, as noted earlier, insulin resistance and its close relative, glycotoxicity (damage from high levels of glucose), represent what is arguably the most common contributor to brain aging and cognitive decline.

Energy creation doesn't simply happen—it's coded for in our DNA, and one of the major sources of code is a gene called PPARGC1A, which tells cells how to make a protein called peroxisome proliferator-activated receptor gamma coactivator 1-alpha, which scientists have blessedly shortened to PGC-1. What's more, we know that exercise—particularly strength training—plays a huge role in the production of PGC-1,[16] and that's a very good thing because, as it so happens, PGC-1 protects cells against oxidative stress, reduces mitochondrial dysfunction, regulates the biochemical signals of inflammation, and improves insulin sensitivity—all effects that you by this point will easily recognize as relevant to the myriad insults that contribute to cognitive decline and neurodegeneration.[17]

As is the case for aerobics, there's no right way to train for strength and there is similarly no easy way to measure the biochemical effects of strength training. But this is also an area in which intuition and proxy measurements give us most, if not all, of the information we need. If you strength-train several times a week—using free weights, machines, or your own body weight—you know when you are getting a good workout versus when you are just going through the motions. You feel it in the moment and in the days to follow. And if in each given exercise you move slowly and safely toward learning what your maximum weight is—or your maximum reps at less than your maximum weight—then you also have a good way to measure change over time.

One important note on those changes: If you're bench-pressing a figurative bus right now, should you always be able to do so? If you lift for years on end, should you eventually be able to lift two buses? Or three? Not necessarily. Our bodies change in many ways over time—as a result of biological aging, sure, but also as a result of becoming *differently fit*. Someone who is 350 pounds is lifting that much weight every day—the muscular machinery needed to support that is quite profound. If that person loses 175 pounds, would I worry that they can no longer squat as much weight as they once did, or do as many reps at a certain weight? Of course not.

To take advantage of the cognitive benefits that result from strength training, I recommend at least three days and preferably four days of strength training each week. (You can split the difference by employing an every-other-day weight training regimen.) For those who are biologically younger, this regimen should be geared toward building and maintaining both muscle mass and strength. Because aging is associated with losses in mass and strength, a condition known as sarcopenia,[18] as we experience aging these exercises should be aimed at slowing those losses.

Many people who are strength training use supplements that support muscle growth, and there is a whole field devoted to supplementation for muscle building and workout stamina. As we discussed in chapter 7, I do believe there is a time and place for supplementation. If you are struggling to build and sustain muscle, you may wish to consider, and perhaps talk to a trainer, about some of the following options:

- Whey protein, which is a good source of protein to build muscle by enhancing the repair of the minor tears that occur during strength training
- Creatine, which supports muscle energetics
- Chromium and amylopectin, which support muscle building, especially in response to a suboptimal dose of whey protein[19]
- Hydroxymethylbutyrate (HMB), a metabolite of the amino acid leucine that, by several mechanisms, including altered cholesterol metabolism, supports muscle growth[20]
- Ursolic acid, which is found in rosemary, sage, and apples, among other sources, and supports fat burning

Not everything that helps build muscle also helps improve cognition and, even if it might, the trade-offs are worth careful consideration. Growth hormone, for instance, can increase muscle mass and strength, but has not been shown to impact cognition in those with decline. Likewise, testosterone has been used for years to increase muscle mass, which it accomplishes both with an anabolic effect (building muscle) and anticatabolic effect (preventing breakdown), and we have supplemented it to reach optimal testosterone levels in those being treated for cognitive decline in our trials. But balance is key: It is important to avoid the side effects associated with supraphysiologic levels of testosterone, such as an increased risk of heart attack, hypertension, insomnia, aggressive behavior, liver damage, and prostate enlargement.

HIGH-INTENSITY INTERVAL TRAINING

The idea of interval training dates back at least to the mid-1900s. So, if you've ever run competitively, you'll likely recall—either with some fondness, some detest, or a little of both—running "intervals" around a track. A common workout is 100 meters at a full sprint, followed by 100 meters of walking, followed by another 100 meters at a full sprint, and so on for a mile. It wasn't until the last decade, though, that the "go hard, break short, go hard again" concept really began to catch on outside of running workouts, as trainers began adding free weights, machines, and bodyweight exercises into the mix. For whatever reason, this was also the point in which the words *high-intensity* were added to *interval training* and the acronym HIIT gained steam. These days the term is hard to avoid. There are physical trainers who specialize in HIIT, HIIT workouts specifically designed for pickleball players, and even HIIT ballet.

Of course, it's not unusual for a fitness trend to catch on quickly only to have people lose interest. We can see this very clearly in the rise and fall of words like *tae bo*, usage of which peaked in 2003; *jazzercise*, which had its heyday in 1986; and *vibrating belt*, a purported weight loss tool that was all the rage in 1969. So, the fact that Google registered a 450 percent increase in the use of the term *high-intensity interval training* between 2010 and 2019 might not mean much on its face. What has fascinated me, however, is just how quickly the health research community has jumped aboard the HIIT train.

HIIT is now the subject of scores of high-quality studies each year.

The Brainspan Workout

Not surprisingly, most of those research efforts have been focused on either holistic health benefits or more specifically on cardiovascular or musculoskeletal health. Increasingly, though, HIIT is being explored for associations with cognitive health, and the data are compelling.

That shouldn't be too surprising. In one way of thinking, HIIT exists at the intersection of aerobic exercise and strength training. It stands to reason that it has many of the benefits of both, and it does. And while I'm not yet convinced that HIIT is a "two-fer" in the sense that it can completely replace more traditional aerobic and weight-assisted exercise, there does appear to be something special that is happening inside the human brain when people engage in short rounds of intense exercise punctuated by short breaks—and my belief, at this time, is that this has a lot to do with hormesis.

The concept of hormesis began in the field of toxicology, where it was coined to refer to the ways in which a low dose of a poison, venom, or other contaminant might invoke a beneficial response on an organism—a tiny bit of herbicide, for instance, sometimes causes a plant to grow faster, bigger, and stronger, whereas a little bit more will do what herbicides are intended to do.[21] The idea has quickly gained popularity with regard to any environmental agent in which a low dose has a beneficial effect and a high dose has a detrimental effect.

People often shorthand this as "what doesn't kill you makes you stronger," an idea popularly espoused by the German philosopher Friedrich Nietzsche, although that's not precisely true. There are plenty of things out there that won't kill you but also won't build you up. But for many stressors, there also appears to be a hormetic point, up to which exposure is beneficial, and a hormetic breaking point, at which it becomes detrimental. Increasingly, scientists are seeing that many stressors, at or below the hormetic point, appear to have holistic antiaging effects.

The most popular and well studied of these effects is fasting—which, as you know from chapter 7, is an important element of a diet aimed at lifelong cognitive health. Obviously, if you withhold food from any plant or animal for long enough, it is going to die. But again and again, energy restriction has been shown to lengthen the lifespans of model organisms—from roundworms to fruit flies to mice to monkeys. It's very hard to do these sorts of experiments on long-lived humans, but using epigenetic patterns and other hallmarks of biological aging, researchers have shown that it is likely that long-term caloric restriction positively affects human aging, too.[22] Some scientists have come to believe that this happens because fasting invokes a survival response—our cells are evolutionarily primed to conserve energy when stress is sensed.[23]

Is HIIT another form of acute stress that invokes a survival response? Well, I think all exercise is. But the short and intense nature of this particular type of exercise might simply be better at hacking into the hormetic response, perhaps because it stimulates our bodies in the sorts of ways that our ancestors were stimulated when they were *truly* in danger. Such dangers were often brief, such as a quick sprint away from a potentially deadly snake. But if our ancestors' bodies produced only enough energy in those moments to respond to a very acute danger, they would have nothing left for the rarer times in which the danger did not immediately subside. Those who evolved to produce extra energy in those times were more likely to survive, and those are the ancestors from whom we descended! This may be why HIIT creates a sudden surge in mitochondrial capacity beyond that which is needed to simply provide the energy being used to fuel that particular exercise,[24] a surplus that helps us maintain the energetic balance that is threatened by aging. This is evidenced by studies that show that HIIT has positive effects on cognitive performance[25] and executive function in old and young people alike.[26]

So, I strongly recommend HIIT (except for those with significant cardiovascular disease). And although I don't think you can always take a shortcut and do a single session of HIIT in place of a purely aerobic session and a separate strength-training session—cutting two workouts down to one—I do think that HIIT can (and should) stand in on a one-for-one basis for either of those other types of exercise, provided that

The Brainspan Workout

you are still regularly getting dedicated aerobic and strength training sessions as well.

BLOOD FLOW RESTRICTION TRAINING

One of the things I appreciate about HIIT is the fact that, as a combinatory exercise strategy that can offer many of the benefits of aerobics and strength training, it has a substantial return on investment in terms of time and effort. It is for this same reason that I have become increasingly enamored with the potential to get "more bang for your buck" through the use of blood flow restriction (BFR) devices, often known as occlusion training and by the brand name KAATSU, which is a company that makes bands that fit snugly around a person's upper arms and legs during exercise, but not too tight that it could drastically cut off blood flow, as an emergency tourniquet is intended to do.

Blood flow restriction training (BFRT) might sound familiar to you as a hormetic intervention—a bit of extra stress to a body that has evolved to need these sorts of shocks to engage its full defensive capacities.[27] Indeed, this is part of the logic that has made it popular with many world-class athletes who have used restriction bands to enhance training or to reduce training time. When I first heard about it, I thought it may be an overhyped exercise fad. But I've also learned over the years that while it's OK to be skeptical, initial suspicions shouldn't prevent you from reading the literature, talking to people about their experiences, and even going so far as to try out things for yourself—especially if there are no safety concerns. So, that's what I did.

BFRT isn't a new idea. This therapeutic practice was developed more than fifty years ago by the Japanese researcher and power lifter Yoshiaki Sato, who designed the KAATSU bands first for his own use, and then for his friends and family members, before the idea began to gain attention in the physical therapy and bodybuilding communities. He later teamed up with scientists from the Yokohama Sports Medical Center to demonstrate that the restriction of blood during exercise prompts an outsized muscle repair response.[28] Put simply: Athletes training with occlusion were getting bigger and stronger muscles while doing almost no extra work! Today the practice is widely used for athletic training, reducing injury recovery time, and preventing muscle atrophy. Indeed, at the 2020 Summer Olympics in Tokyo (held in 2021), many athletes were using KAATSU bands, and this greatly pleased Sato, who lives nearby in the city of Fuchu. "It was always just a matter of time," he told the *New York Times* as the games got underway. "I just did not think it would take this long."[29]

It may take even more time before one of the most intriguing aspects of BFRT—its potential impacts on brain health—will also reach the mainstream. I do suspect that will happen, though, and I'm not alone.

Intrigued by what they'd learned in collaboration with Sato, the researchers from Yokohama Sports Medical Center asked the next logical question: Why? To begin moving toward an answer to that question, they collected plasma samples from a small number of study participants before and after a series of leg extension exercises, then compared what they'd found. Later, they did the same thing, with the same participants, who were this time wearing occlusion bands, and discovered that growth hormone, norepinephrine, and lactic acid all showed marked increases when blood flow was restricted.[30] These findings set off a flurry of other research, and in the years to come other research teams demonstrated that BFRT appeared to also have an impact on insulin growth factor, vascular endothelial growth factor, and hypoxia-inducible factor—all of which are parts of signaling pathways that have been associated with neuroplasticity and cognition. Taking note of all these findings in 2018, a team of German scientists from Otto von Guericke University, the German Center for Neurodegenerative Diseases, and the Center for Behavioral Brain Sciences called for greater focus on the potential impact of occlusion training on cognition.[31]

In 2021, an international team of researchers reported that BFR improved executive function even when used in association with very mild exercises, like walking.[32] In that same year, a group headed up by neuroscientists in Iran reported that BFRT had a significant impact of people's sleep quality, moods, and performance on a test that is commonly used to check for cognitive impairment.[33] Studies like these, and several others, led a team of Spanish researchers to conclude that occlusion training was a promising strategy for cognitive protection. "Blood flow restriction therapy seems to be beneficial in neurological disorders without adverse effects," the team wrote. But, they cautioned, "it will be necessary to conduct clinical trials that use larger sample sizes and greater homogeneity . . . in order to be able to more objectively compare effectiveness compared to other treatments, clinical studies are needed."

Therein lies the rub. Large clinical studies of the sort the Spanish research team correctly wished to see are very expensive, while BFR bands are relatively inexpensive (unlike a drug, for example) and last a long time. Thus, once again, researchers who are interested in these promising associations face barriers to funding.

And so, yes, it's fair to say that my enthusiasm for BFR is based on limited clinical evidence, but it's worth noting that it's also based on a quickly accumulating body of patient outcomes.

Take Michelle, for instance. The seventy-two-year-old retired nurse and lifelong skier came to us with what appeared to be a quickly deteriorating memory—and also two very damaged knees and one bad shoulder, which made exercise difficult. But, of course, exercise is an essential element of the ReCODE Protocol and, as you now know, strength training plays a big role. With KAATSU bands, though, Michelle was able to lift very light weights with greater benefit. And while this was just one of more than a dozen interventions that she was prescribed—and it can't be said to have been the deciding factor in her recovery—it did seem to bring her great joy.

"I was an athlete all my life," she said. "To not be able to do much exercise over the past few years, and to watch my muscles really deteriorate as a result, was really making me sad. I think that even if my memory hadn't been going so quickly, this alone might have made me depressed. But with the bands, you know, I can do pretty easy exercises, and I know I'm getting a result, and in fact I can see a result when I look in the mirror."

And it's not just how her body looked. It's how her brain responded! Over the next year, Michelle completed an inspiring cognitive turnaround—with no reported memory loss or any other signs of cognitive decline.

Of course, as I've repeated many times in this book, one need not wait until their body and brain are deteriorating to begin taking advantage of the holistic benefits of exercising with occlusion bands. And while this form of exercise assistance is not for everyone—those with uncontrolled hypertension, chest pain, heart failure resulting in an inability to exercise, and sickle cell anemia, among other conditions, should avoid restricting blood flow. Everyone should, of course, consult their personal physician. The barriers to entry here are so minor and the potential benefits so substantial that I believe more healthy people will engage in this sort of training.

Indeed, I suggest that it should be a regular element of everyone's strength training workouts, used several times a week during lighter exercises (at a load of about 20 to 40 percent of your max) for up to fifteen minutes. (Don't go longer than that. Remember, the hormetic rule is: A little bit is good; a lot is not.)

EXERCISE WITH OXYGEN THERAPY

The human brain is a remarkably complicated thing, and anyone who tells you they truly understand it most certainly does not. And yet there are some things that can be simplified to very basic ideas. One is that our brains need oxygen to survive.

When the brain is deprived of oxygen—even for just a minute or two—it creates a cascade of pathophysiological crises. Our mitochondria cease functioning and our energy plummets. Our neurons die. After three or four minutes, the damage is often irreversible. After five or six minutes, death is almost always imminent, and there are very few people who have ever gone more than ten minutes without oxygen and lived to tell the tale.

Fortunately, we live in a world of plentiful oxygen. It's literally just a breath away! Because of this, our ancestors didn't need to figure out how to store it for a long time in their brain cells. We evolved to get it just when we need it.[34]

But because our atmosphere is only about 21 percent oxygen, that's as much as we could ever get—up until the point that we learned how to isolate it, store it, and administer it in higher concentrations. Today, pure oxygen—especially that administered at higher-than-atmospheric pressures—is a well-established treatment for infections, anemia, burns, decompression sickness, and radiation injuries, and has been shown to be a promising therapy for the reduction of inflammation, a reduction in free radicals, and the activation of neural stem cells.[35] Studies have also indicated that concentrated and pressurized oxygen may offer substantial improvement in cognitive function for patients suffering from mild cognitive impairment, Alzheimer's disease, and vascular dementia. (Reduced energetics due to vascular compromise is an important contributor to brain aging, and there are over one million Americans with vascular dementia.)[36]

It might seem reasonable to assume, then, that we may be able to combine the benefits of exercise and supplemental oxygen to impact brain health. Alas, research in this area is nascent. (This is yet another case in which there is no drug to sell, ergo there isn't a huge profit impetus for understanding this potential association.) But there are some strong indications that this O_2 combo punch may indeed be a powerful way to improve cognitive function. For example, in one study of people suffering from Long COVID—a common symptom of which is severe and lasting brain fog that presents much like a dementia—a rehabilitation program using exercise with oxygen therapy (EWOT) resulted in MoCA score changes from a range indicative of mild cognitive impairment to levels that were nearly normal, in just six weeks.[37]

Data like those have been intriguing for my dear friend Julie G.

After employing the ReCODE Protocol to recover from a substantial and frightening brush with cognitive decline in her late forties, Julie has become an ardent evangelist for lifelong brain health. She also understands that the protocol is a work in progress, and she has enthusiastically embraced opportunities to try interventions that may help others reverse—and better yet, prevent—cognitive decline and neurodegeneration.

To that end, in 2024 Julie purchased one of the many home EWOT systems that have recently become available and increasingly affordable for people who want to experiment with combining exercise and supplemental oxygen. She began using the system—which includes an oxygen concentrator, an oxygen bag, and a specially fitted mask—during her HIIT exercise sessions, in which she interspersed walking and sprinting. And because Julie is a bit of a data junkie, she also decided to do a pre- and post-exercise check of her arterial stiffness using a tool that measures aortic pulse wave velocity. "I've been using it on and off for the past five years," Julie later reported, "but have never been able to budge the numbers too far, even with pretty extreme exercise." But in her very first use of EWOT with HIIT, Julie's arterial stiffness score improved by 36 percent. "I had never seen that big of a drop with any exercise or other intervention," she wrote.[38]

That's a single data point in a very nonclinical setting, so you may take it for what it's worth, but as more ReCODE Protocol patients have begun using EWOT, and I've heard more and more stories of success, my confidence in this intervention has only grown—not only as a way to turn back cognitive decline but as a way to offer a small and needed boost to everyone with a brain that is working "on the edge."

And that's all of us! As you'll recall from the uphill bicycling analogy that we've used several times in this book, our brain is a system that is running very near its limits at all times, and thus is vulnerable to any little thing that goes wrong. This includes the efficient processing of oxygen to fuel the brain's ravenous energy needs in a world in which, even when we are exercising at our max capacity, the best we can get is

21 percent oxygen. The good news is that it doesn't take a big bump in either the concentration or pressure of oxygen to help set things right. To extend the analogy, consider what the advent of electric bikes has done for people with uphill commutes. The "pedal assist" feature doesn't actually add a ton of power. It just offers a little something extra. That's what I have come to believe EWOT is doing, too.

It's fair to note that EWOT can be cumbersome. It takes time to set up. You need to keep the system clean. And exercising with a mask on—and a tube connecting that mask to the oxygen bag and concentrator—can feel very awkward. Although home EWOT systems are indeed becoming more common and more affordable, this is not a purchase that most people can make without some consideration of the cost versus the benefit. For this reason, if it is possible to do so, I recommend finding a fitness or exercise therapy center where you can try it out a few times before buying. For me, it's worth it, but these are considerations everyone should make in association with what feels right to them (and, as I will always hasten to say, in consultation with their personal physician). If you do decide that EWOT is right for you, I recommend getting in at least one session each week during an aerobic or HIIT workout.

It is important to point out that some EWOT systems feature cycling out of hyperoxia and into hypoxia (lower oxygen than room air), then back into hyperoxia. This is yet another contributor to hormesis, and in this case it leads to increased secretion of trophic factors. However, there are some caveats to be aware of. First, as we've noted many times in this book, reduced energetics are almost always a common contributor for anyone who is already suffering from any degree of cognitive decline, so reducing support even further can be concerning. If you do wish to add EWOT to your recovery regimen, you may wish to consider waiting at least six months, and only after you have begun to see improvement from other interventions, before you try it. Second, simply cycling from hyperoxia to normoxia may be enough to achieve the same trophic support, since it is a relative hypoxia. Finally, there has been little published research so far showing that this hypoxia approach is helpful for those with cognitive decline. For those who are in prevention mode, without any decline, there is much less concern about EWOT systems that induce brief cycles of hypoxia.

JUST GO EXERCISE

Many years ago, a patient named Leena came to me with a request that I have since heard many times from others.

"Can you give me a list of the exercises I should do?" she asked. "I'd really like to make sure I'm doing the ones that will have the most benefit."

I understood where Leena was coming from. Patients are used to doctors prescribing specific medicines. Why wouldn't they expect us to prescribe specific exercises?

I do try to be accommodating. I tell people that they should be doing aerobics and strength training several times each week—and it's not a bad idea to get the benefits of both of those by doing a HIIT workout once in a while. Beyond that, I suggest blood flow restriction training and exercise with oxygen therapy. And lately, inspired by the work of Ryan Glatt, a trainer and brain health coach at the Pacific Neuroscience Institute who combines fitness with problem-solving in a "brain gym," I've started suggesting combining exercise with some fine motor skills exercises, cerebration (thinking—as you might do while learning ballroom dancing, for example), or both. Ryan, as an avid video gamer, found significant health benefits and greater fitness when he discovered Dance Dance Revolution, in which players must move rapidly with their bodies to match quickly changing musical and visual cues. He now helps his clients find activities that likewise activate their minds and bodies in ways that feel fun.

Still, some people want to know exactly how far to run each day or how many miles to ride on their bicycle. They want to know what exercises they need to do, at what weights, at what number of sets and reps.

I get it, I do. But here's what I'd like them to understand most of all:

The Brainspan Workout

The best exercise for your brain is the exercise that you will do, consistently, right now. And that is the exercise you enjoy.

It's running if you love to run. It's bicycling if you enjoy that. It's rowing on a river in the crisp morning. It's lifting weights in your backyard with the sun on your back. It's soccer or pickleball or basketball or ballet or tae kwon do. It's whatever you enjoy, because if you don't enjoy it, you're far less likely to do it.[39]

So, first and foremost, if you wish to keep your brain very healthy—now and for many decades to come—what I want most of all for you is to find an exercise regimen that you really do enjoy doing. We can build from there.

Let's assume it's golf that you love, "a good walk ruined" as they say. That might not at first sound like a great vehicle for aerobics, strength training, HIIT, BFR, or EWOT. And that's true, it's not, but it's actually a great start—especially if you have any interest in being a better golfer. That's because, as it turns out, aerobic exercise and strength training have been shown to improve recreational golfers' games.[40] HIIT workouts, meanwhile, have been shown to have a positive impact on golfer's drives.[41] BFRT has been demonstrated to help amateur golfers improve their game.[42] And EWOT has been shown to improve athletic recovery.[43] So, yes, it's true that golf itself might not give you the exercise you need to assure a hundred-year brainspan, but if you love that game and you wish to play, play well, and play for a long time, it's the perfect start.

What I told Leena back then, and what I've told many patients since, is that I want you to find your perfect start. Build up your exercise regimen from there. That's the pathway to an ageless brain.

9

CLEANSE AND RESTORE

> Practice does not make perfect. It is practice, followed by a night of sleep, that leads to perfection.
>
> —MATTHEW WALKER

For years, Sally had figured that six hours of sleep each night was adequate.

As a nursing professor, she understood that six hours was a bit on the short side of ideal, but she also knew that it wasn't a tremendous outlier for people in her profession. After all, the medical world is awash with folks who encourage their patients to get seven or eight hours of sleep each night, even though they often do not do so themselves. She was also not an outlier among her peers; according to the National Center for Health Statistics, about a third of adults between the ages of forty-five and sixty-four get fewer than seven hours of sleep a night.[1] What's more, Sally felt pretty well rested.

Then her brain began to rebel.

It started with twitching, as if her hands were hiccupping. Neurologists know these kinds of involuntary muscle contractions as myoclonus, which is very common in early-onset and rapidly progressive neurodegenerative disease,[2] and Sally was aware of this, too, but she did her best to ignore it.

Next, she began using the wrong words to describe things she knew perfectly well. This is known as aphasia, and it is also a common side effect of neurodegeneration,[3] but such are the powerfully soothing whispers of the demons of denial that Sally set aside the possibility this might also be a sign something was wrong with her brain.

It wasn't until Sally forgot to pick up her grandchildren from school—twice in a single month—that she recognized that something might really be wrong. That's when denial gave way to panic. And, alas, for good reason. By the time she had found her way to me (after having a positive amyloid PET scan, finding out she had a single copy of ApoE4, and worsening during a clinical drug trial), her MoCA score was indicative of mild cognitive decline, the third of four stages of Alzheimer's, and her cognoscopy showed dozens of heightened risk factors for neurodegeneration.

Although I could not promise her that we could "fix" her broken mind, I sought to reassure her that her cause was not hopeless. "I know it might not seem like it," I told her, "but to me, these results are good in one very important way. If you had the same symptoms yet none of these results were positive, then we wouldn't know what contributors to address. But what we have here is a road map."

That map began with addressing the six hours of sleep Sally was getting each night. I reassured her that it wasn't a bad starting place, but the goals I encourage my patients to patients aim for include:

- At least 7 hours of sleep each night
- No more than 8.5 hours (more than 9 is associated with risk for dementia)
- REM sleep of at least 1.5 hours
- Deep sleep of at least 1 hour

- Oxygen saturation (SpO$_2$) while sleeping of at least 92 percent, and preferably over 94 percent
- No evidence of sleep apnea, thus the goal is an AHI (apnea/hypopnea index) of 5 or lower

We'll delve deeper into each of these recommendations in this chapter.

Like every other step that is part of the ReCODE Protocol, better sleep won't change everything—and, indeed, sleep isn't the only thing Sally needed to work on to reclaim a more youthful brain—but it's no secret that rest is absolutely vital to healthy cognition, and I don't know of anyone who has suffered from cognitive decline who has made a full recovery without addressing this pervasive insult. (The person I often call Patient Zero, whom I saw in April 2012, had been getting four to five hours of sleep per night. She corrected this and other contributors, and she has now sustained improvement for over a dozen years.)

Of course, the point of this book is to help you *prevent* the development of cognitive decline. That means you'll only need to do what Patient Zero did (and, with apologies for the spoiler, what Sally did, too) if, for any reason, the prevention failed.

But I do recognize that it can be hard to convince people to change their sleep habits. While about a third of adults sleep for *fewer* than seven hours a night, that means two-thirds sleep for *more* than that. So, on paper, anyone who is getting those seven or eight hours is meeting recommendations, and that leads a lot of people to assume they're getting exactly what they need. But brain-sustaining rest is more complex and comprehensive than that. How many hours you get each night is an important place to start, but it's only just that—a start. The kind of rest that protects a youthfully functional brain is just as much about what happens before those hours, during those hours, and after those hours. It's about the breaks your brain is getting throughout the day. It's about the way you approach the world.

That's why it's important to think of rest not only as something that happens at the *end* of the day, but as something our brains must do when we need to counteract the cognitive insults we accumulate *throughout* the day. And when we consider sleep in this way, it becomes clear that there are actually two things we can do to make rest more productive. One is to get better sleep. The other is often ignored: to prevent the insults, so that our sleep is more effective once it comes. It is when we focus on both sides of this coin that our brains are best protected from aging. So, in this chapter, I'm going to start by talking about what we can do to sleep in a way that minimizes brain aging, but then I'm going to focus on some of the best ways to prevent the insults to begin with.

THEY'RE GONNA WASH AWAY

There's a beautiful song by the singer and songwriter Joe Purdy, the first track on an album called *Julie Blue*, that almost perfectly encapsulates the role of sleep in our lives. In "Wash Away," Purdy sings:

I got troubles oh, but not today
Cause they're gonna wash away

Purdy may not have realized it, but he was singing about a formula for a healthy body and brain. While we often and rightfully worry about the damages that can be inflicted upon our bodies and brains by toxins and contagions, and while it's clear that physical injuries can have lasting consequences, one of the least appreciated day-to-day insults to our brains comes by way of the many stressors we are exposed to from the time we wake to the time we sleep.

If you're regularly jarred awake by the wailing of your alarm clock, you've started your day with an accelerated stress response, which is made worse if your sleep has been less restorative than you need.[4] If you're a parent, you know that raising children is often an exercise in managing stress and, sure enough, researchers studying working mothers found a striking connection between parenting stress and elevated levels of the stress hormone cortisol, especially on weekday mornings.[5] If you commute to work on a freeway—even if you don't experience discourteous drivers but especially when you do—your stress response is elevated even more.[6] Do you feel calm, relaxed, and joyful at work? (If you do, you're fortunate, as occupational stress and resultant burnout is a very common and, it would appear, growing problem.[7])

From morning to night, we're awash in "troubles," as Purdy sings, and this triggers the production of stress hormones, which accelerates the accumulation and deposition of protein and byproducts in the brain that are hallmarks of cognitive impairment. But what if, with each new day, we could somehow "wash away" the insults of the previous twenty-four hours? If we could, our brains would effectively be as healthy today as yesterday, the day before, and the day before that. Those days would add up to weeks and months. Those weeks and months would add up to years and decades.

There is a system that serves this function in much of the body. As you might know, the lymphatic system washes away excess waste, moving these byproducts of metabolism where they need to go to be eliminated through breath, perspiration, bowel movements, and urination. It's long been known, however, that the central nervous system doesn't have any lymphatic machinery, so it wasn't clear how, or whether, the brain was able to clear waste products in a similar way—getting them away from our neurons and circulating in the rest of the body to *then* be cleared out by the lymphatic system.

Now we know that the brain does indeed have an analog, the glymphatic clearance pathway—a key element of which wasn't discovered until 2023 when researchers announced the identification of a superthin and delicate cellular membrane that surrounds the brain and acts as a platform for immune cells to monitor the movement of cerebrospinal fluid for signs of infection and inflammation.[8]

Just consider this for a moment: Humans have been studying the brain for millennia. We've had sophisticated imaging tools for decades. And yet this structure, the subarachnoid lymphatic-like membrane, has gone unnoticed until just the past few years! It's discoveries like this that

Cleanse and Restore

make me shake my head when people say things like "brain aging is inevitable." If we're still discovering what the brain is and what it does, how can we possibly conclude that anything is inevitable?

In the few short years since the discovery of the glymphatic system and the structure that permits much of its functioning, we've learned so much. For instance, we have discovered that as we sleep, the toxic proteins that have built up from the day's onslaught of insults are washed away through this filter. It has also become clear that glymphatic function is inhibited by adrenergic tone—in other words, when your sympathetic system is firing, whether because of stress or threat or watching a scary movie or whatever, your glymphatics don't do their job properly.

It might seem commonsensical, then, that the number of hours of sleep matter. The more sleep we get, the more time this filter has to work its gunk-cleaning magic. That's partially true, but what the emerging research also suggests is that *deep sleep* is a particularly powerful glymphatic activator. The human brain doesn't fall into and stay in a state of deep sleep but rather reaches this state of rejuvenative slumber multiple times in a night (often early in the night, which is why the old rhyme "early to bed, early to rise . . ." is not a bad idea) as it moves through a pattern of variations in the electrical signals in our brains that looks a bit like a classic sine wave, with multiple descents into deep sleep and multiple ascents into REM sleep, the state most associated with dreaming and that is characterized by rapid eye movement. REM tends to occur later in the night and early in the morning. So, total sleep time matters not simply because the filter is on for longer but also because it is permitted to run at its maximum power more times over the course of a night. Thus, one of the best things we can do to help the glymphatic clearance pathway do its work is to optimize our slumber with the express intention of reaching a sufficient dose of deep sleep each night.

Before I get into that, I need to make an admission: On the night before I wrote these words, I slept quite poorly. Normally, I try to get at least one hour of deep sleep and one and a half hours of REM sleep. This is both what is recommended as part of the ReCODE Protocol and what I suggest to anyone who is trying to prevent brain aging, and there are now very accurate sleep trackers—from rings to watches to bedside monitoring devices—that allow us to understand both how we sleep in a single night and how our sleep is trending over time. It's the latter of those two ways of thinking of sleep that I hope you'll focus on most, because it's simply a fact of life that there will be times when we don't get the rest we want and need. For me, last night, it was almost certainly because I didn't get everything done during the day that I wanted to, so instead of winding down as my bedtime approached, I was still in my study working, thus I got a late start on my sleep and this resulted in a shorter window of opportunity to achieve the sort of sleep I need. Indeed, when I woke up, I learned that I missed both the deep sleep and REM targets, owing to less total sleep time and fewer complete cycles through these important stages. What this means is that my glymphatic system didn't get a chance to "run at maximum" as many times as usual. This was not cause for panic, though. It's true that it is preferable to have *no* nights like this, but what I'm concerned about in my life—and what I hope you will be concerned about in yours—is positive movement toward more deep sleep and REM over time, and then general consistency once you get to the one-hour and one-and-a-half-hour targets for deep sleep and REM sleep, respectively.

And that's why I'm going to tell you something that very few sleep experts would say publicly: How you get there matters less than the fact that you do, in fact, get there. This is to say that while "sleep hygiene" has correctly become the focus of many of the books and podcasts about more healthful sleep, I often meet people who fret over not being able to adhere to the most common advice for creating the "perfect" conditions for slumber, chiefly because—at least at that point in their journey toward better sleep—they feel that the thing being advised would actually have a negative effect on their sleep.

By way of example, I recently met with a patient named Barbara who had the habit of turning on a police scanner as she went to bed each night. "It's something I started doing after my husband died," she said. "I know it would probably make people feel anxious hearing the dispatchers describing the various accidents and crimes that are happening all around them, but it soothes me—it reminds me that there are people out there whose job it is to protect me, and it eases my worries."

This absolutely defies one of the fundamental principles of sleep hygiene—which is that noise is bad for sleep and thus bad for health.[9] And if we had been unable to improve Barbara's sleep in other ways, it would absolutely have been one of the things I'd want to focus on. But after addressing several other issues that were likely to be impacting her sleep, Barbara's sleep tracker indicated she was achieving about one and a half hours each of deep and REM sleep. Her numbers are good and holding steady. Why mess with success?

So, while I firmly believe that the principles I will be sharing in the next few pages will help improve your sleep, I don't want you to get hung up on one thing that seems to be a nonstarter or, worse, might be counterproductive. Try a different tactic first. The goal isn't to get everyone to sleep in the exact same way, after all. The goal is meaningful glymphatic activation.

Turn Off the Screens and Wi-Fi
If there's one thing that almost everyone can do to get to sleep easier each night—and one thing that almost everyone is resistant to—it's changing our relationships to the screens that surround us.

Televisions, laptops, and smartphones are literally designed to keep our brains passively engaged—to give us just enough stimulation to keep us docile but not to put us to sleep. Part of this is the action that's happening on the screen, which is detected by our eyes and travels to our brain's visual cortex by way of the visual thalamus (a structure called the lateral geniculate nucleus), which is an intricately interconnected region that has the capacity to quickly engage virtually all the other parts of the brain. Another part of it is the content, especially that which is loud, emotionally arousing, or thought-provoking, that can stimulate stress, anxiety, or interest. Still another part of it is the light emanating from the screens, especially that which exists in the blue spectrum, that has been shown to suppress the production of melatonin, the hormone typically released as the sun goes down to cause us to feel sleepy. The complex nature of why screens stimulate us to stay awake defies attempts to "hack the system," for instance by wearing blue light–filtering glasses or watching only soothing content. Many people do find that using a filter to avoid the stimulating high-energy part of the spectrum for the last few hours before bed is helpful. And switching to dark mode on your laptop in the evening can help, too. But these tactics alone are generally not enough, because it's not just one element of the experience that is preventing sleep. Truly, the most powerful play here is to eliminate all screens from your bedroom and to stop using devices at least an hour before you intend to sleep.

Lights Off

We cannot be certain when the first brain evolved, but fossils of an ancestor to both arthropods and vertebrates, called kerygmachela—which looks to me a little like a cross between a modern centipede and a hickory tree branch—suggest that organs resembling those that exist in most modern animals' heads are at least 520 million years old.[10] Mammalian brains came along about 300 million years after that, and the human brain, very close to its modern form, showed up just about 100,000 years ago.[11] The world has changed a lot during all these stages of evolution but, at every stage of development, one thing remained very constant: a twenty-plus-hour day (Earth days were a bit shorter hundreds of millions of years ago) that was marked by distinct periods of light and darkness. By way of contrast, electric light is about 200 years old and didn't reach the majority of human civilization until the past half century. Put simply, we're *deeply* evolved to respond physically and mentally to the rising and setting of the sun—and not at all evolved to deal with an ersatz sun that can be turned off and on at any time of the day or night. It should come as no surprise, then, that electric lights

Cleanse and Restore

wreak havoc on our circadian rhythms, chiefly by preventing sleep.[12] To make matters worse, John Marshall, an ophthalmologist at Moorfields Eye Hospital at University College London, is among those who have warned that the long-lasting LED lights that have become so popular in recent years may be a problem, rather than a panacea, since they emit more blue light than the old incandescent bulbs.[13]

Now, I am not so crazy as to suggest that we should all be living in perfect synchronicity with the sun without any assistive electrical lighting, but from the perspective of brain health, every step in that direction is a step in the right direction. And the first step—one that almost everyone can take, even if they are initially reluctant to do so—is to either dim your house lights or begin turning them off one by one over the course of a few hours as you get closer to bedtime. Better sleep is virtually guaranteed as a result!

Bedtime and Wake Time

Because we are so deeply wired to respond physiologically to a twenty-four-hour circadian cycle, in which the sunrise and sunset might come at different times throughout a year but change very little from one day to the next, better sleep can often be achieved by adhering to a consistent sleep cycle, with a set bedtime (no later than midnight, although a few hours before is better) and set wake time. This is the best way to create the seven-to-nine-hour window that most people need to achieve the requisite cycles of deep sleep that engage the glymphatic system, but it can be very hard for many people. That includes me; of all the principles I recommend to prevent brain aging, this is the one I struggle with most.

It can be especially tempting to sleep in on the weekends or during holidays, but this is a practice that makes consistent sleep more difficult on adjoining days. A 10 p.m. bedtime and 6 a.m. wakeup each day (give or take one hour on either side), maintained with consistency every day, will result in better sleep for virtually everyone.

Sleep Apnea

From the time we are in utero, humans can perceive sound.[14] While some noise from the outside world passes through the amniotic fluid and bodily tissues that envelop the womb, most of the sounds in this space emanate from the mother.

Her voice, sometimes. A sneeze here and a cough there. And, given the frequent discomforts of pregnancy, more than a few groans, no doubt. But the soundscape is dominated by two specific sounds: the beating of the mother's heart and the inhalation and exhalation of her breath. There is the lub-dub-lub-dub-lub-dub-lub-dub emanating from the opening and closing of the heart's valves. And then there is the slower haaa-fwaa-haaa-fwaa-haaa-fwaa as the lungs expand to bring in air and contract to push it away. Before we are aware of anything else, we are being conditioned for a world of rhythm!

This may be why so many of us enjoy placing our head on a loved one's chest to listen to their heart and breath. This may also be why so many of us fall into a euphoric trance when running at a sustained pace. Perhaps this is even why we love rhythmic music and dance.

But is this also why we have memory? I'd never considered that until I read a study by a team of neuroscientists from Germany and the United Kingdom that demonstrated that the well-known association between sleep apnea and inefficient consolidation of short-term memories into long-term memories is at least in part a result of apneic interruption of our aerobic rhythms.

In this study, which was led by Ludwig Maximilian University of Munich cognitive neuropsychologist Thomas Schreiner, researchers showed study participants 120 images, each of which was associated with a specific verb. For instance, the participant might be shown a picture of an apple and given the verb *bite* or they might see a picture of people getting onto an airplane and be offered the word *travel*. The fascinating thing about the way verb-object or verb-scene pairs like this allow researchers to test memory is that the association is both related, meaning it is likely to create a robust neural connection, and not necessarily obvious, you can also *eat* an apple or *slice* an apple, just as you can *fly* on a plane or *board* a plane. After being introduced to these image-word pairs, the subjects went to sleep in a lab where their breathing was monitored. When they awoke, they were shown the images and quizzed on the associated words. And when the researchers analyzed the data, they discovered that those whose breathing was most rhythmically consistent were most often the ones who remembered the associations.[15]

Of course they were! We literally come to life in a world of rhythm. It makes perfect sense that our own rhythms—including the primary sounds that we can hear as we sleep, our own breathing—would serve as a pacemaker for the process of memory consolidation.

There are many other and perhaps even more obvious reasons (such as low levels of blood oxygen, hypertension, obesity, cardiovascular disease, gastroesophageal reflux [GERD], esophageal damage, and genetics) why sleep apnea—the repeated stopping and starting of breathing throughout the night, which impacts about one in five people[16]—is such a neurological scourge, with demonstrated impacts on rates of Parkinson's,[17] Alzheimer's,[18] and all-cause cognitive decline or dementia.[19] Indeed, sleep apnea is one of the common underlying conditions we see in patients with cognitive decline, and one of the most commonly missed diagnoses, with only about 20 percent of cases diagnosed,[20] despite the fact that the most common symptoms—loud snoring, waking up during the night gasping for breath, awakening in the morning with a dry mouth, headaches, daytime sleepiness, and trouble focusing during the day—are quite palpable.

Most often, the cognitive effects of sleep apnea are attributed to the impacts of intermittent hypoxia that accompany the frequent awakenings and breathing pauses that people with this condition experience, but the sleep-damaging effects of adrenaline surges may also contribute. As oxygen is the primary catalyst for the redox energetics that fuel our brains, it is no surprise that when it is in short supply, neural damage is virtually inevitable. But regardless of which of these insults is causing the greatest cognitive damage, the upstream source is the same: interrupted breathing during sleep.

The most common treatment for this is the continuous positive airway pressure machine—a CPAP device, which helps maintain open airways during sleep, thus reducing or eliminating the episodes of hypoxia and sleep disruption. Studies have demonstrated that consistent nightly CPAP use can lead to improvements in memory as well as in overall cognitive performance,[21] thanks to the restoration of normal sleep rhythms, airflow, and the restorative functions of deep and REM sleep.

Alas, many people find CPAP machines cumbersome, and actually prohibitive to the goal of good, restorative sleep (one patient said he would prefer death to CPAP). There are alternatives, including oral devices that hold a user's tongue in place or slide their jaws forward to keep their airways open, positional therapy devices that help people adopt the sorts of positions most conducive to unobstructing their airways, and—generally as last resort—surgical procedures that alter the geography of a person's mouth and throat to widen the upper airway, or retainer-like devices that achieve the same effect. Another very successful approach has been weight loss and reduction in inflammation. A study published in 2024 in the *New England Journal of Medicine* suggested that weight loss from tirzepatide, commonly known by the brand name Mounjaro, may also reduce obstructive sleep apnea.[22] To know which of these options is right for you, a sleep study is often necessary. This is the gold standard for assessment, and while these studies are often performed in sleep labs (which is an attribute that can actually prevent good sleep, since nobody really *wants* to sleep in a lab), home-based sleep studies are increasingly available and often more affordable.

Like so many of the other insults that we have discussed in this book, however, the problem with many of these options is that they may come far too late, after damage has been done to the structures of our brains. I'll reiterate here, though, that the brain is a marvelously resilient organ and we have had success improving cognition in people whose neural machinery has been damaged by *decades* of sleep apnea. But even though this is just one of many insults that impacts the longevity of our brainspans, it should not be ignored. The sooner it is addressed, the sooner we can move on to mitigating other insults. Thus, it is unwise to wait for symptoms. Proactivity is key!

To this end, it is imperative that we monitor our sleep for the signs of obstruction so that we can take action when problems are developing, not once they are firmly established. In addition to sleep trackers, continuous pulse oximeters (which record blood oxygen levels throughout the night and create a report that can be saved for later reference) should be used at least once in a while. This doubles your chances of identifying problems early on. It is so much easier to use a minor intervention like a mouth device than it is to wait until a major intervention like surgery is needed. What's more, these tools provide opportunities to improve our sleep even when sleep apnea isn't the problem.

After I began monitoring my sleep, for instance, using a smartwatch, I noticed that although I try to maintain consistent bedtime and wake times, I almost always got more cycles and deeper periods of deep sleep on Friday nights. I suspect that this is because I feel more relaxed at the start of the weekend. While I do often work on weekends, I rarely have anything that I need to get a jump on first thing on Saturday morning. As a response to these data and their presumed causes, I have sought to limit my earliest meetings on weekday mornings as well.

Of course, I hope to never find out what life with sleep apnea is like, but if I do begin to suffer the early symptoms, I'll be in a position to intervene quickly and decisively. And that, I have come to believe, is a strategy that is absolutely vital for anyone who wishes to achieve a brainspan of a hundred years or more.

Exercise

Since there is a chapter dedicated to exercise, I won't belabor the point of how important it is to cognitive health, save to note the very specific benefits to sleep that I have seen again and again in my patients—and which have been borne out again and again by researchers. Simply put: Increased exercise improves sleep outcomes by better regulating sleep hormones; increasing one's energy expenditure, which leads to more restful sleep; and reducing stress and anxiety.[23]

It should be noted, however, that exercise also raises adrenaline, the hormone that signals our bodies to increase blood circulation, bring in more oxygen, and more quickly metabolize carbohydrates to prepare our muscles for exertion. Exercise is, after all, a mimetic for the sorts of fight-or-flight circumstances our ancestors sometimes found themselves in back when we were not nearly the dominant species that we are today. In those days, anyone who survived an encounter with a beast early in the evening would almost certainly be dead by morning if they did not stay awake and alert through that night. Put another way, in that situation, those who had an adrenaline response that could override sleep signals at night had a survival advantage. That's the response we've inherited, and so it makes little sense to exercise right before bedtime. I tell my patients that morning exercise is best. A noontime run (especially in lieu of a large lunch) is brilliant. A trip to the gym right after work is probably OK, too. But within three or four hours of bedtime, the advantages of physical activity are likely outweighed by the detriments of spiked adrenaline and the resultant trouble getting to sleep or staying asleep.

Food

If you are following a mildly ketogenic, plant-rich, highly nutritive diet like KetoFLEX 12/3—with a nightly fasting of no less than twelve hours, including three hours before bed—you're already doing virtually everything you need to prevent eating habits from interrupting your sleep. But that "three hours before bed" part of the equation is what I wish to emphasize here, because even if you're eating healthy foods, eating right before sleep can lead to discomfort and indigestion (since, when you lie down, gravity no longer helps to keep the contents of your stomach down, which can lead to heartburn or gastroesophageal reflux) and thus impact your ability to sleep comfortably. It can also shift the focus of our brain. Instead of sending hormones to aid in sleeping, our brain is providing the hormones that aid in digestion. As a result, insulin is high, which leads to restlessness and an inability to reach the multiple cycles of REM and deep sleep. And as you now know, our brains need those sleep cycles to wash away the insults of the day.

All of this and more is why I tell patients that it is important to protect that three-hour window of fasting before bed, which permits sleep to do what sleep is supposed to do for our brains, by way of the glymphatic system, and also across the rest of our bodies through autophagy, the cellular recycling process that is triggered by fasting and plays a vital role in cell survival and maintenance.[24]

Alas, this is one of the most common "speedbumps" that patients experience when adopting a lifestyle conducive to restored cognition and longer brainspans. Many of us are conditioned from childhood to eat when we feel even the slightest pang of hunger. Sometimes we feel this impulse to eat after dinner, when we're settled down and sedentary—often sitting in front of a television or computer screen—and the feeling becomes hard to ignore. One way to flip the script on this experience is to make sure you are indeed eating at times when you might not feel so hungry—much earlier in the day. Another is by ensuring your final meal of the day is satiating, especially by including plenty of fiber, which digests slowly in the gut. Spending time reading, meditating, or in conversation with family or friends can also offer sufficient distraction. If all else fails and these feelings of wanting food are impairing your sleep, an ultralow-calorie snack—a small amount of edamame, a few florets of broccoli or cauliflower, a carrot or celery stick, a few nuts, or a "fat bomb" like a shot of extra virgin olive oil—can often offer a degree of satiation without imperiling sleep. These also help to smoothen your glycemic curve and reduce the chance of your awakening at 3 or 4 a.m. due to hypoglycemia.

In the end, though, sustained movement toward a three-hour window that is a "true fast" is quite important, and the sooner you're able to make this a lifelong habit, the better off your brain will be.

Sleep Medications and Supplements

I am not an antipharmaceutical crusader.

Sometimes, because I have been so publicly critical of the utter failures of Alzheimer's medications, people assume otherwise, but it's absolutely not true. There is a time for drugs. There is a place for drugs.

So, for people who need help getting back into rhythm, and either have not been successful in other ways or are in crisis, I am not vehemently opposed to sleep medications that are prescribed by a doctor, taken under supervision, and include a plan for getting off the medication once sleep has improved. But medication isn't a cure. It's a stopgap. And when it comes to sleep meds, it's not even a particularly great one. Most people who take sleep aids get about a half hour boost in sleep, but one of the many trade-offs is the potential for a disruption of REM sleep and deep sleep,[25] which may be why we're starting to see many people turning away from prescription sleep aids after decades of steady increases in the use of these medications.[26]

For these reasons, I strongly recommend that, if a person has adopted good sleep hygiene habits but is still struggling to meet the sorts of sleep goals that will lead to longer brainspans, the first step before turning to pharmaceuticals should be neuroprotective supplements like melatonin, magnesium, and in some cases L-tryptophan.

A steady rise in melatonin use over recent years might be one of the key reasons why sleep medication use has declined, particularly in the United States where melatonin is classified as a dietary supplement and available without a prescription. Our bodies naturally produce melatonin to encourage us to rest at night and help us stay alert during the day. But since dietary supplements aren't strictly regulated and labels might not accurately describe what's in the bottle, it is of utmost importance to purchase brands that have a long and excellent track record and are transparent about their sources and practices. (See chapter 7 and our discussion of nutritional supplements.) I also suggest that melatonin be used only for short-term mitigation of sleep troubles, as long-term use has been less adequately studied.[27] Even in short-term use, though, less is more: Most people do well with 0.5 milligrams to 3 milligrams. And, as is the case for all supplements, it should be truly supplementary to other interventions!

If not for the presence of magnesium in our bodies—which helps regulate muscle and nerve function, blood sugar levels, and blood pressure—we would likely be "on edge" all the time. Thus, many people find they relax easier and get to sleep quicker when they take magnesium before bed. Alas, magnesium is inexpensive at pennies per dose, so there is little impetus for long-term study. But as Colleen Lance, medical director for the Sleep Disorders Center at Cleveland Clinic Hillcrest Hospital in Ohio, said in 2021, "I tell patients you can give it a try and see if it helps. It may not help, but it's probably not going to hurt."[28] In my experience, patients often do subjectively feel better about the quality of sleep they get when supplementing with magnesium, which might additionally explain all the cognitive benefits that come with nutritional intake of magnesium, as we discussed in chapter 7.

Tryptophan is most famously associated with the sleepiness many people feel after eating a big turkey dinner at Thanksgiving. In reality, the *big* part of that is far more impactful than the *turkey* part; there's really not enough tryptophan in turkey to make us feel sleepy. But it is true that tryptophan is converted by our bodies into serotonin, which helps control mood and sleep, and this is why many people find taking supplemental tryptophan beneficial for rest. In most cases, we can get plenty of this essential amino acid from a brainspan diet, as discussed in chapter 7, and foods like SMASH fish, eggs, seeds, and nuts can be good sources.

I have typically recommended supplemental tryptophan to people who are awakening at night and—once again—always in association with other interventions to help improve sleep duration and quality. But I have seen enough evidence in the experiences of my own patients to convince me that tryptophan can help people stay asleep. This observation conforms to a meta-analysis of four smaller studies.[29]

Electromagnetic Fields

If you had told me just a few years ago that you were worried about the ways in which "invisible energy waves" emitting from power lines, cellphones, home appliances, Wi-Fi routers, and computers were impacting your brain health, I'd have asked you to show me your tinfoil hat. Electromagnetic fields (EMFs) are a natural part of our world, produced by the buildup of electric charges in the atmosphere by electrical storms and trapped by the Earth's own magnetic field, but there hadn't been much compelling evidence showing that low-frequency sources of electric activity pose a danger to human health.

But a few things happened in recent years that made me rethink my dismissal of this increasingly common concern. First, EMFs became an increasing focus of the work of Neil Nathan, whose work on biotoxins and chemical sensitivity (which we'll discuss further in chapter 11) has helped many patients with cognitive decline. Dr. Nathan's concerns were not necessarily for the general population, but for a group he has termed "sensitive patients," those for whom—for whatever reason—insults that don't affect most people can cause severe distress. This was particularly interesting to me in the context of increased neurodegeneration within a species whose brains had come to run "on the edge," and whose suffering from diseases like Alzheimer's had begun a precipitous fall from that cliff.

Next, I became familiar with some studies—small ones, to be sure, but compelling nonetheless—that indicated statistically significant effects in sleep quality for people exposed at work to extremely low-frequency EMFs at high-voltage substations,[30] and improvements in sleep for individuals whose sleep environments were designed to buffer the EMFs in everyday technology.[31]

Finally, I listened to many patients, reflected on what they were telling me, and checked myself. I am frequently critical of physicians

Cleanse and Restore

who don't listen to their patients, and some of mine—people I'd been working with for years—had become compelled by emerging data to reduce their EMF exposure by making sure that everything that could easily be unplugged in their home was indeed unplugged when they weren't using it. These are, after all, patients on the edge, and they felt as though unplugging in this way was helpful.

When I started thinking of EMFs in these ways, I began to recognize something quite important: Anything that can knock people off the edge, when they arrive at that precarious place, can also bump them toward the edge as life goes by. As Dr. Nathan has pointed out in his book *The Sensitive Patient's Healing Guide*, EMF sensitivity is often part of a syndrome of multiple sensitivities (for example, to various chemicals) and is typically initiated by exposure to mycotoxins or the bacterium *Bartonella*. The syndrome is mediated by three processes: limbic system alteration (affecting emotional and behavioral response), vagal nerve transmission (reduction in the body's rest and digestion response), and the activation of the allergic-response white blood cells known as mast cells. Together, these represent a protective "fight or flight or faint" response to stress.

So, as skeptical as I was initially about sensitivity to EMFs, I could no longer ignore the pattern that emerged. The sheer number of people who have reported sensitivity to EMFs demands we take this possibility seriously. Furthermore, there are deep parallels between the "connection to protection" mode switch that we have observed in people with advanced brain aging and disease, and a similar switch that the bodies and brains of sensitive patients simply can't turn off when exposed to insults that others might not even notice.

The practical problem currently is that, while we can measure EMFs, there is no clinical test that will tell us whether someone is indeed suffering from EMF exposure. I hope that such a test will be available in the not-too-distant future. Meanwhile, data are accumulating that at least some people may indeed be sensitive to EMFs, so turning off your Wi-Fi at night and keeping your distance from other sources of EMFs is likely to be a good idea for long-term brain protection.

KEEPING CLEAN

From France came the torch of the Statue of Liberty, still ten years away from being raised over Bedloe's Island in Upper New York Bay. From Germany there were two heavy cannons, known as Krupp guns. Mexico sent paintings from the landscape artist José María Velasco and the neoclassicist Santiago Rebull. Japan sent a display of miniaturized trees.

And at the Centennial Exposition in Philadelphia in 1876, the nearly 10 million visitors to the global exhibition of art, technology, and progress were treated to marvels of American ingenuity as well—from a novel monorail locomotive that connected two of the busiest exhibition halls to a machine that could produce 100,000 screws in a single day to a 650-ton engine that powered the entire fair! But perhaps no invention provoked more excitement than a powered washing machine that had been sent to the exposition by the Shakers, from a community in Canterbury, New Hampshire. Driven by a reciprocating arm, which could be powered by water or steam by way of a drive pulley, the machine's wash tubs slid back and forth on metal tracks, removing dirt by the violent movement of the clothing against the washtubs and the churning of the soapy water inside.

There was one setting for this machine. In the years to come, as washing machines became more prevalent in homes and better at the chore they were designed to do, simply washing clothes wasn't enough. Eventually there were settings for hot, warm, and cold washes, and combinations thereof for wash and rinse cycles. There would be choices for the size of the load, the types of fabrics to be washed, the colors of the clothes, and the dirtiness of the job—and it is that last set of choices that, for whatever reason, came to my mind as I sought to explain why simply getting good sleep isn't always enough to "wash away" the insults that accumulate in our brains each day. For even today, with our high-tech washing machines, some of which literally employ artificial intelligence, and our specially designed soaps that are the product

Cleanse and Restore

of thousands of chemists and other scientists locked into an "appliance science" race, sometimes grass stains still get the better of us.

Of course, the best way to prevent a grass stain is to not slide on the grass. But as anyone who has ever removed such a stain can tell you, the second-best way is to address the problem immediately: to prevent the stain from setting.

If the glymphatic system is our washing machine, and stress is our dirt, then the actions we take to reduce and address stress throughout the day are the difference in the settings we need on our machines to make sure everything comes out clean. And since we all have trouble sleeping sometimes—indeed I've yet to meet a "perfect sleeper"—it's important that we all have strategies to reduce and address the stress that accumulates throughout the day, as soon after it happens as possible.

Meditation

I'm a little hesitant to admit that, as a data-driven neuroscientist, I may have initially misjudged the neurocognitive benefits of meditation. In my defense, back when I was training to be a doctor, very few people regarded meditation as a medical intervention. It was a way to be at peace, to discover connection, to seek wisdom, or to connect with one's spiritual beliefs. All of that was well and fine by me. But a cure for diseases of the brain? What's more, a way to prevent these diseases? No, that wasn't something I would have even considered to be possible.

I was wrong. The data on meditation's potential benefits for cognition are compelling, and I have come to believe that much of this is related to its ability to help people prevent, or quickly address, stresses that would otherwise materialize as part of the daily accumulation of toxic proteins that the glymphatic system needs to wash away each night as we sleep. In dozens of trials with thousands of participants, meditation has been demonstrated to alleviate anxiety, reduce depression, and even mitigate physical pain.[32] Relative to one of the challenges we have mentioned about food and sleep, meditation may help control appetite.[33] And unlike exercise, it can be done in the hours before sleep without adversely affecting sleep quality. Indeed, it has been shown in many studies to improve the quality of sleep,[34] so it might be thought of as a twofold intervention for brain aging—one that can both prevent "stains" from accumulating and improve the glymphatic wash cycle to boot. It shouldn't be surprising, then, that research on people at virtually every age, from many cultures, has shown cognitive benefits suggesting that a lifelong practice of meditation may help prevent age-related cognitive decline.[35]

I'm not the only scientist who had initial doubts about the power of meditation to improve cognition. So did neuroscientist Sara W. Lazar, who found her way to a yoga studio in Boston after she injured herself training for that city's famous marathon.

"The yoga teacher made all sorts of claims," Lazar said in 2018. "And I'd think, 'Yeah, yeah, yeah, I'm here to stretch.' But I started noticing that I was calmer. I was better able to handle more difficult situations. I was more compassionate and open hearted, and able to see things from others' points of view."[36]

Lazar would go on to become a researcher at Massachusetts General Hospital and professor of psychology at Harvard Medical School, and she is now one of the world's most renowned experts on the connections between meditation and cognition. Among her early studies, in 2005, was one that suggested long-term meditators, those who have dedicated at least seven years of their lives to the practice, had substantially thicker brain regions associated with attention, interoception, and sensory processing.[37] That finding was based on just twenty study participants—it's not unreasonable to assume that the differences that Lazar and her team found in the brains of long-term meditators could have been caused by many different aspects of their lifestyles. But for twenty years to come, study after study from her labs, and many others, have affirmed that initial finding: Meditation is a force multiplier for brain health across our lifespans.

While today I understand the neurological *impacts* of meditation quite well, I am no expert on the practice itself, and I'm not convinced that there is a "best" way to meditate. Lazar came to these habits by way of yoga, which may have beneficial effects on cognitive functioning through improved sleep, mood, and neural connectivity, according to work from a team led by Gretchen A. Brenes and Suzanne C. Danhauer, both of whom specialize in the connections between health and yoga practices at Wake Forest University.[38] Much of Lazar's work has focused on the practice of mindfulness, an awareness of the present without judgment or cognitive elaboration. I've had many patients who have found meditative success through transcendental meditation, a practice of detachment often guided by the repetition of a mantra or a sound.[39] I've had others who swear by tai chi, the Chinese martial art known for its slow, intentional movements, often called "moving meditation."[40] Forest bathing, a Japanese practice known as shinrin-yoku, in which people simply spend time being calm and quiet in nature, has gained widespread popularity in recent years as a meditative practice with profound psychological benefits.[41] All of these meditative and quasi-meditative practices have been demonstrated to reduce stress and improve sleep. All are associated with substantial long-term benefits in cognitive health. The recurrent substance of all these practices is having a regular practice of stress relief. This makes our waking hours more restful and our sleep more restorative.

RESTED AND RESTORED

Sally has now done very well, with improved cognition for seven years. However, after six years she had a temporary setback, and when the potential contributors were identified, severe sleep apnea was one of three issues we found (along with one infection and a new toxin exposure). When these were treated, she achieved her highest cognitive test scores yet. She has been a model for surviving and thriving after an initial bout with cognitive decline, and she is among the many patients who have sought to help others *avoid* coming anywhere close to the precipice of cognitive decline.

The glymphatic clearance pathway is one of the keys to restorative sleep and reaching the goal of making dementia a rare problem. Sleep, in turn, is key to the ability of this marvelous system to wash away the culminating stresses of each day. And habits focused on lowering or immediately addressing those stresses—from identifying a personally fulfilling meditative practice to finding an effective therapist—are key to ensuring that we don't have to always get the very best sleep for the glymphatic system to do its work.

It bears noting that stressing about stress doesn't help. Losing sleep over lost sleep doesn't help.

You've got troubles, sure, but on this day, know they can be washed away.

10

THE BRAIN'S FLEX FACTOR

You can't teach an old dog new tricks.

—JOHN HEYWOOD

Oh yes you can.

—FROM THE RESEARCH OF KAVLI PRIZE
WINNER MIKE MERZENICH

Even after all these years of working with patients who are suffering from cognitive decline, I still cannot fathom the depth of the anxiety and anguish they and their families go through.

To make matters worse, this is a time of great upheaval in medical practice, so patients who seek help from so-called "centers of excellence" are often treated with ineffective drugs and the standard, hopeless mantras that we've been telling people for generations.

"There is nothing that will prevent, delay, or reverse Alzheimer's," they are told. "Alzheimer's currently has no cure and no survivors."

It is confusing and disheartening for patients and their families to be given such dire messages—especially when they learn that there are several thousand patients being treated by a few hundred practitioners who would disagree with those outdated claims. That confusion is compounded when they read about peer-reviewed studies that repeatedly show cognitive improvements, such as the reports my colleagues and I published in the *Journal of Alzheimer's Disease* in 2022[1] and 2023.[2] This can have a whiplash effect on patients, who come to me and other doctors who are trained in the ReCODE Protocol hoping for incredible turnarounds—and in some cases, *expecting* miracles. I have seen some recoveries that have astounded even me but, alas, these are not results I can promise.

It is thus understandable that, sometimes, family members' frustration boils over. This is most common when it comes to people who have started treatment in the late stages of dementia—which is why I keep harping on the recommendation that everyone should get evaluated and begin active prevention or earliest treatment starting in their thirties, long before dementia sets in. Frustration is also common among those who are early into the protocol, or who have not been willing or able to complete more than a few parts and whose family members are upset that there has not been a rapid turnaround.

I truly wish it were that easy. Someday, perhaps, it will be. Until then, alas, there will be cases like the one I faced with a young man named David.

David's father, Douglas, was a seventy-eight-year-old former contractor deeply afflicted by dementia—unable to care for himself much beyond the very basics of showering, brushing his teeth, and putting on his own clothes. Douglas could no longer cook; he would leave things burning on the stove. He could no longer find his way to the grocery store or church, both of which were just a few minutes' walk away from his home; he would get lost and end up miles away. He couldn't even hold much of a conversation anymore; after a minute or two his eyes would lose focus, and he would become detached from the discussion.

We immediately began working through the ReCODE fundamentals. We studied Douglas's diet and recommended some immediate and long-term changes. We advised him to add some moderate exercise to his routine. We monitored his sleep schedule and began making adjustments there as well. But David wasn't happy with his father's immediate progress and wanted us to do more.

For years, he had heard commercials for so-called "brain training" programs that could purportedly protect against cognitive impairment, dementia, and Alzheimer's disease through the use of specially designed games and puzzles aimed at challenging and stimulating different parts of the brain. (Unfortunately, a lot of those claims were overblown, and one of the highest-profile companies advertising those programs ended up paying a $2 million settlement after the Federal Trade Commission charged the organization with making unfounded claims.[3]) David had also heard—correctly, as it happens—that there had been studies on some of these programs that did indicate some improved memory and attention in older adults.[4] A little additional homework told him that this sort of training was part of a personalized therapeutic program that my research group had employed on ten patients, nine of whom had begun showing improvement in cognition within six months.[5]

"If it was good enough for them, why wouldn't it be good enough for my dad?" he asked me.

That was a fair question. Could the ReCODE Protocol be helpful for someone so far along in cognitive decline?

Brain-training programs generally take the form of puzzles, memory exercises, and attention tasks that are aimed at improving a wide range of cognitive functions. The basic idea is that just as physical exercise strengthens our muscles, respiratory, and cardiovascular systems, brain training may strengthen our cognition by stimulating neural pathways, creating new connections to replace those broken down by the accumulated insults of aging.

I'd first become aware of the theories that underpin these programs in the mid-1980s when I heard a neuroscientist named Mike Merzenich speak at a conference in Asilomar, California. His talk blew me away—he showed that the sensory cortex could actually expand anatomically with stimulation. This was completely unexpected in 1984!

Yet it fits perfectly with what we see today when our patients with cognitive decline increase their hippocampal, temporal lobe, or parietal lobe volume.

For more than a decade, at that point, Mike had been studying the ways in which our brains work to make sense of new signals when the old machinery has broken down in some way—a vital attribute of neuroplasticity.

I am so thankful to Mike for opening the eyes of so many of us to the remarkable potential for neuroplasticity throughout life, and even for those with Alzheimer's. Up to that point, after all, scientists had believed that neuroplasticity was a feature of our brains that ended after childhood. If information was water, the pipes had all been laid. So, while you could still increase water pressure, and you could feasibly change the kinds of liquids running through those pipes, there was an upper maximum for flow. And once the plumbing broke down, that was that. Or so we thought.

But using what were then state-of-the-art microelectrode mapping tools to demonstrate which very specific neural connections are activated when humans and other mammals do something or learn something, Mike was showing that, if and when it is necessary, we can create new pipes, both temporary and permanent, to handle different flows, separate different liquids, and—most important of all—replace pipes that have broken down.

By way of example, at that time Mike had recently demonstrated that monkeys that have lost the capacity to use some of their fingers could reappropriate the cortical connections that had previously been dedicated to those digits for other functions.[6] For the entirety of my medical training, I had been taught that certain areas of the brain are associated with certain features and activities of the body—a cortical "homunculus" that trapped us into a world in which, if a specific part of the brain was damaged, the result was permanent disability in whatever structure that part controlled. It was this way of thinking, in part, that led us to the horrific conclusion that cognitive decline was inevitable

The Brain's Flex Factor

and irreversible. On that day in California, though, Mike was essentially saying "to hell with that."

But knowing that the brain can reorient itself and knowing how to force it to do so in a beneficial way are two different things, and Mike and others have dedicated decades of research to that latter pursuit. Among short-term studies of brain-training programs, the data are encouraging but mixed, with some studies showing small positive gains and others showing no changes in test groups beyond what was seen in controls. Long-term studies are, of course, much harder to fund and administer and are therefore rare. But one caught my imagination—and has buoyed my hopes that researchers are on the right path. The Advanced Cognitive Training for Independent and Vital Elderly (ACTIVE) study was a randomized, controlled trial of more than 2,800 initially healthy older adults that examined three brain-training programs over the course of ten years. At the end of that decade, those who had engaged in a training program aimed at improving the speed at which the brain receives, processes, and responds to information had a 29 percent lower risk for dementia than those in the control group—and those who received even more of this speed-of-processing training had even greater rates of resilience against this form of neurodegeneration.[7] Importantly, some of the other programs that were tested in the ACTIVE study did not produce such jaw-dropping results, indicating that even well-designed brain-training programs aren't always that effective. It's thus reasonable to be skeptical of groups that claim that their interventions will have these sorts of results if they have not been vetted in the same way. And, alas, the prevailing evidence strongly suggests that while some people who have already experienced substantial neurodegeneration may improve with the help of a well-designed brain training program, the efficacy across the board isn't great for individuals in that situation.

I've come to believe this is because once neurodegeneration has begun, the added cognitive demands of training may simply further tax a distressed brain, unless the support is increased and/or the demand reduced. These are, after all, people who are fighting to stay above water. Adding cognitively challenging tasks may essentially be tying an anchor to their ankles. Furthermore, there is a degree of stress associated with failing any test and, as we've previously discussed, stress is a common contributor to cognitive decline.

And this brings me back to David and Douglas. What I told David at the time was that brain training, and other forms of cerebral stimulation such as light therapy, may be helpful but should be eased in, since brain training is very much like intense physical exercise. You don't want to start as a weightlifter if you're completely malnourished. This is why we start with removing the identified insults (such as a damaging diet, sleep apnea, and toxin exposure) and supporting the brain (with metabolic flexibility), then ease into therapeutic brain training—at which point someone like Douglas would be in a better position to form new neural connections.

As you have undoubtedly noticed, many of the principles of this book suggest that what is good for helping people who are suffering from the accumulation of insults that come with aging is also good for protecting those of us who haven't reached that state of crisis, and never want to. Similarly, people suffering from cognitive impairment are often more likely to reach a point of suffering sooner and decline faster. In our proof-of-concept clinical trial, despite the fact that all the subjects were in the third or early fourth stages of cognitive decline, MCI, or early dementia (the same stages of patients in antiamyloid antibody trials that demonstrated no cognitive improvement, only modest slowing of decline), every single one of them improved their scores on BrainHQ, which is the brain-training program developed by Mike Merzenich.[8]

Some have argued that brain training only improves one's ability to take the brain-training tests and does not spill over into daily cognition. But in our trial, 84 percent of the study participants improved both their cognitive scores and their symptoms.

This is only one of the many reasons why I believe Dr. Merzenich's demonstration of lifelong neuroplasticity is a foundational principle for all of us who would like to improve our cognitive performance today and reach one hundred with that function perfectly intact.

EVERY DAY, MONTH, AND YEAR

While Dr. Merzenich and others have taught us that our brains can indeed remain malleable throughout life, the reason why the opposite idea persisted for so long is because our lives aren't often organized in ways that are conducive to *using* that capacity.

Consider, for instance, what you do most mornings of your life. Do you wake up in the same bed, in the same room, in the same home, in the same town? Do you shower in the same place, using the same soap, drying with the same towels? If you begin your day with a cup of coffee, does it come from the same cupboard, is it brewed in the same machine, and do you drink it in the same spot as you do most other mornings, or get it from the same coffee shop on your way to work? And speaking of work, do you get there in the same way, using the same sidewalks or streets, in the same kinds of vehicles, seeing the same scenes pass you by as you commute?

There's nothing inherently *wrong* with any of this. It's how most of us live and, in fact, our lives would be chaotic if they weren't organized in these sorts of ways. But such environments require very little neuroplasticity because, for the most part, the connections that were created in our brains years and even decades ago are all we need to survive and even thrive for a very long time.

This is why, in many cases, we begin to notice our struggles only when our environment changes. Once, a patient told me that she only began to worry about her cognitive health after her company changed locations. Almost a year after the move, she still found herself driving to her old office on occasion. Another patient explained to me that he felt furious after his church moved to a fully digital system for tithing. He knew his was an irrational emotional response, but he just couldn't get over the fact that, after decades of donating by check, he was going to have to learn how to do it a new way.

We used to think of situations like this as being related to people being "absent-minded" or "set in their ways." We now know that there is a neurological explanation: They've formed deep and lasting connections in their brains for certain functions of life and, over the years, have gotten inadequate practice at forming *new* connections from day to day, month to month, and year to year.

Brain-training programs may be one way to get "plastic practice." I believe that well-designed programs such as BrainHQ, which we have employed as part of the ReCODE Protocol, have the potential to reverse or prevent cognitive decline, it also makes a lot of sense to organize our lives in such a way that we are consistently creating new neural pathways just by virtue of the ways in which we live. Now, we obviously can't change *everything* about our lives every day, but I don't think we need to. Instead, we should be striving to take on a small new cognitive challenge each day, a medium new cognitive challenge each month, and a big new cognitive challenge each year.

What does a small daily cognitive challenge look like? Well, almost anything that requires you to engage your memory, attention, language, perception, problem-solving, or decision-making in a way that encourages you to do something differently than you've done before. So, if you're a regular crossword puzzler, it's not just a new crossword but a new type of puzzle altogether—a cryptogram, or a solitaire game, or a word search, or a sudoku. If you go to the same coffee shop each morning, choose a shop you've never been to before (and perhaps order a different drink than you've ever had before). If you work from the same desk at home each day, it's temporarily setting up your office at a different location and perhaps getting started and finished at a different time. None of these need to become new habits and, in fact, the point is that they aren't something that continues day after day—they simply interrupt normalcy to provide your brain with an opportunity to create new connections. Even the act of coming up with 365 different ways to change your daily routine is an exercise in plasticity-supportive cognition, but it may be hard at first to make the breaking of habits into a habit, so keeping a list of ideas for these daily challenges and then journaling what you will do or have done each day is a good way to track your progress and success. Once you've completed a specific daily cognitive challenge, is it off-limits for the rest of your life? Certainly not! In fact, it might be very beneficial to go back to a task you haven't done for a while to reengage or re-create a synaptic connection that was built for the same purpose.

A medium monthly cognitive challenge might be reading a few books from a very different genre of literature than you're used to, learning the rules to and playing a new game or sport, or cooking meals from an ethnic tradition that is new to you. Much like the daily challenges, the key here is to vary not just the activity but the type of activity from month to month. Thus, one month your cognitive challenge might be learning the basics of cooking Indian food but the next month you shouldn't shift to Japanese food; you should learn to play backgammon. The following month you shouldn't jump to mahjong; maybe you should start listening to jazz. Certainly, if you come to love making sushi or become a big fan of Miles Davis, there's nothing wrong with making these new affections part of your life in the long term. But the intent here isn't expertise; it's cognitive experience.

An annual cognitive challenge is an obviously bigger commitment, and this is why it makes sense to align these efforts to grander ambitions. If a trip to Florence is on the horizon, an Italian language course may be in order. If you've recently become fascinated by chess grandmaster and social media influencer Hikaru Nakamura, perhaps a year dedicated to the study of the game of kings is right for you. True mastery of any subject after just a year of effort is unlikely, but a year of work on anything—even in increments of just ten or fifteen minutes each day—is virtually guaranteed to result in a better-than-average capacity to understand and engage in that activity for life.

I hasten to note my belief that none of this should be directly connected to your professional path or pursuit at which you have already developed substantial expertise. If you're already an avid golfer, improving your short game is a laudable goal but unlikely to offer neuroplastic gains on par with taking up oil painting or learning to play the guitar. Plasticity is truly about interrupting the old with the new.

In these daily, monthly, and yearly increments, our brains are developing the capacity to reorient in the short, medium, and long term. And thus, when we are ultimately confronted by changes to our lives in these same sorts of temporal spans, we have the cognitive capacity to adjust with ease.

OTHER FORMS OF PLASTICITY-INDUCING STIMULATION

It always bears noting that mice are not humans—an obvious statement that somehow seems to get lost when people are speaking about promising medical research. So, when I recently saw that a team of scientists from Australia and China had identified a therapy that "rescues functional synaptic plasticity after stroke," the very first thing I checked was the animal involved.

It was, indeed, a mouse.

That doesn't make the work insignificant. Strokes are among the most debilitating events that can befall a human being—one that can result in a brain that looks as though it has been through decades of aging, particularly when it comes to the damaged capacity for plasticity. But results in animal models need to be approached carefully. To that end, it was the specifics of their therapy—the use of light flashing at a speed that is thought to be similar to the rate at which our brain cells communicate—that excited me, because this finding aligns with human studies that have shown that rhythmic light stimulation at this same rate, 40 hertz, facilitates connectivity across cortical regions in healthy human brains.[9] Researchers at MIT, meanwhile, have demonstrated that both light and sound transmitted at this rate, which is in the gamma range, may be associated with some neurological and behavioral benefits among people with Alzheimer's disease, although their work was limited by its brief duration and small sample size.[10]

Thus, the mouse models had helped to advance something that is much harder and costlier to study in humans: the range of potential impact that 40-hertz sensory stimulation might have, exercising neuroplasticity across a spectrum of brain pathologies. This is already becoming popular as part of the overall strategy for preventing plasticity loss in humans from an early age.

I'm encouraged by research like this—because while 40-hertz therapies seem to be a very promising pathway toward rebuilding (and, better yet, maintaining) synaptic plasticity, it's not the *only* sensory stimulation pathway.

Microcurrent electrical neuromuscular stimulation, in which weak electrical currents are sent into the body through electrodes placed on the skin, may activate plasticity, a technique that is being explored to suppress tinnitus (ringing in the ears), which appears to occur as a result of the altered processing of auditory signals and rerouting of information when synaptic connections falter as a result of aging.[11] Transcranial magnetic stimulation, which uses an alternating magnetic field to target electrical signals in specific areas of the brain, is also being explored as a way to incite synaptic plasticity, although the research in this area remains nascent.[12] Photobiomodulation, the use of specific spectrums of light to stimulate physiological responses, has been shown to improve synaptic plasticity in mice as well.[13] I think there is great promise in all these modalities.

Promise and payoff are, of course, two very different things. It may be that, like so many of the interventions we have discussed, what is effective for one person won't be as effective for another. But I predict that one or more of these brain stimulation modalities will factor into mainstream brain care very soon, because the potential harms seem negligible if not nonexistent. And I hasten to note, once again, that you shouldn't wait to experience profound cognitive decline before engaging in interventions that interest you. No matter what your age, it may be wise to add some form of brain stimulation to a wellness regimen aimed at maintaining exceptional cognitive function right now.

PROTECTING COGNITION WITH CONNECTION

Fundamentally, neuroplasticity enables us to alter our behavior for competitive advantage. Constant change—seasons, interactions, goals, physiology—require adaptation moment to moment. And few things support neuroplasticity like the constant interactions that come as a part of maintaining human relationships.

Everyone changes over time, after all, and this may be one of the underappreciated reasons why having meaningful interpersonal relationships is so very good for protecting our brains against aging. The plasticity of our brains is supported by the plasticity of *their* brains, and evolution is endless. Whether it is a spouse, sibling, friend, or cherished neighbor, everyone changes over time. They will learn new things, take on new traits, make new habits (both good and bad), and develop new opinions (some of which we might come to agree with and others of which we assuredly won't). This is a tremendously powerful environment for neuroplasticity because, once these relationships have been established, we are hesitant to simply let them go. We have some brain cells in the game, so to speak, so it behooves us to adjust our understanding and expectations over time—to engage in the acts of cognitive exercise that are required to sustain these relationships. Indeed, this may help explain why the loss of a spouse is such a powerful trigger for all manner of brain pathologies,[14] and why researchers have found a substantially increased risk of dementia for married people whose spouses are diagnosed with dementia[15] (although it could also be the shared environmental exposures). When the most clear and present source of modulation of our own day-to-day patterns is no longer playing that role, the cognitive challenges we need to ensure neuroplastic capacity quickly dwindle away.

So, relationships are of clear importance to an ageless brain—they are, in a very unromantic way of thinking, helping to train us to remain plastic as we age—but this again is not something we can simply apply as a salve to people who are already experiencing cognitive decline. Indeed, once a person is suffering in this way, it is infinitely harder to build the sorts of deep and abiding relationships that would actually meet this need. Thus, learning to be a good partner, friend, and neighbor—that is to say, one who understands, accepts, and even encourages the changes that their friends and loved ones go through across their lifespans—is a cognitively protective act. Parents who are not teaching these skills (and alas, many, it seems, are not) are doing their children a grave disservice. Schools that are not enforcing these values (and alas, many, it seems, are not) are unwittingly contributing to the growing epidemic of cognitive decline. Perhaps most perniciously, media sources that tell people how to think and encourage them to believe that people who are not exactly like them are wrong (or even "evil") are robbing their consumers of the cognitive challenge of adjusting their ideas and expectations of others over time. That may be why highly partisan people (on any part of the political spectrum) are more likely to suffer from faulty memories, struggle with perception of reality, and difficulty distinguishing fact from fiction.[16] The influential economist Noreena Hertz has argued that loneliness is one of the most common traits among conspiracy-minded far-right thinkers in the United States, United Kingdom, and France.[17]

Ironically, I suppose, if you are feeling lonely, you are not alone. Loneliness is on the rise globally, and it doesn't just impact our brains when we are in our final years. Researchers have demonstrated that these feelings are likely to negatively affect executive functioning and attention even among people in their early twenties.[18] It thus behooves us at any age to fight against the social, technological, and political forces conspiring to disconnect us from others. Unfortunately, this is not really a book on making friends and finding life partners, but there are several experts in this field whose work I highly recommend, including the Christian theologian Jennie Allen, the author of *Find Your People: Building Deep Community in a Lonely World*,[19] and the secular psychologist Marisa G. Franco, who wrote *Platonic: How the Science of Attachment Can Help You Make—and Keep—Friends*.[20] These, to me, are not simply books about connection; whether intentionally or not, they are books about protecting long and healthy brainspans.

THE SOCIAL BRAIN

Sadly, long-term relationships aren't always within our control. Partners and siblings pass; friends and neighbors move away. Some of us are simply less fortunate than others in this regard. So, these sorts of relationships should not make up our only form of social interaction. In fact, sometimes having a few deep and durable relationships might make it *less* likely that we will engage in the other vitally important form of cognitively protective social interaction—the random connections we can make throughout the day *if* we interact with people we either don't know or don't know very well.

I emphasize the conditional conjunction *if* because there is good evidence that humans are becoming more insular—less likely to say hello to a stranger, attend civic events in their communities, or engage with an acquaintance in a way that makes that person a little less of a stranger and a little more of a friend. In some respects, this was a predictable consequence of the first truly global pandemic of the twenty-first century. This was a time in which just about everyone was keeping at least a little more social distance between themselves and people they didn't know. Truth be known, though, we were moving toward less sociability with strangers long before COVID-19 came along. Many of us were already studying online, working online, and dating online. Some of us were already having groceries delivered instead of going to the store ourselves. And even when we ventured out into the world to do so, plenty of us were already checking out all by ourselves without a human cashier to greet with a few words about the weather. However, one of the associations we have noted since beginning the ReCODE Protocol is that those who participate in a support group as part of their overall protocol tend to improve more often than those who do not.

However we got here, we are, as the author Andy Field has beautifully reflected in his book *Encounterism*, missing out on the "sites of friction and possibility, of anxiety and joy. There is a spontaneity to an encounter that demands we approach it with a degree of uncertainty and vulnerability, a softness that leaves space for compassion and empa-

The Brain's Flex Factor

thy in a world where both can often be in short supply. An encounter is an opportunity."[21] These opportunities are not simply chances to connect to another human being; they are moments in which our neuronal circuitry must rearrange itself to process a new face, hear a new voice, and decide whether some new information is valuable enough to be stored for longer-term use. These are the moments in which plasticity is practiced.

The cognitive benefits are pretty much immediate, according to research led by University of Michigan psychologist Oscar Ybarra, who found that even brief episodes of social contact with strangers have a positive impact on memory, focus, and emotional regulation.[22] But the benefits of making this kind of socialization a habit are almost certainly going to be reflected in our brainspans as well, particularly in an era in which the immediate back-and-forth conversations that once defined almost all human communication have been replaced by discussions that "have moved online or on to phones, where you can control the conversation: you can take all the time you want to respond," Ybarra said in 2021. "This is the first time in history where you don't have to respond immediately to someone who's saying something to you."[23]

The gravity of this shift becomes quite clear when we consider what happens in a human brain during a conversation. As the interaction begins, for instance, sensory information, especially auditory signals, enters our brains and is processed by the respective sensory areas. Neurons in the auditory cortex then translate these signals into basic sound units, which are transmitted to other parts of the brain for further processing via neurotransmitters across synapses. That's when the brain's language centers become active as a rapid sequence of synaptic transmissions decode syntax, semantics, and context. Of course, a conversation is not only about listening, it's also about knowing how to respond, which means the prefrontal cortex is also involved as we decide what to say and when to say it in sequence with the other speaker. And since language is more than verbal, the limbic system helps process emotional clues from the other participant and dictates emotion-signaling responses. All the while, our synapses are strengthening in some parts and weakening in others as our brains process a flood of new information, using memory of past events and episodic projections of potential futures to filter the most potentially useful bits away from the potentially useless bytes.

We can get all this exercise from a polite moment of small talk with a stranger on the subway or a colleague in the elevator—but only if we *choose* to do so.

THE EMOTIONAL BRAIN

Any sort of brain training that is sufficient to prevent cognitive decline is necessarily both neurological *and* psychological. This is to say that it cannot simply create a brain more capable of attention, it must create a brain more capable of affection. It cannot just produce a brain that is better at accessing memory, it must make a brain that is better at controlling mood. It cannot only make us better at decision-making, it must make us better at dealing with others. A well-trained brain—one that is getting sufficient plastic practice—is one that is psychologically resilient.

This means getting to know—and tolerate—the full spectrum of our emotional experience as human beings. There are, by some accounts, tens of thousands of distinct emotions, or perhaps some smaller number with distinct temperatures—the difference, for instance, between annoyance and ire and anger and rage.[24] However you reckon it, there will be times in your life that you will experience an emotional state that is unusual in its textures and complexities. Perhaps for you that might be envy or awe. For someone else it might be pensiveness or ecstasy. Sat with for a few moments, experienced as more than a fleeting thought, these emotions are opportunities to exercise changing neuronal connections, just like talking to a stranger at a ballgame or getting to know the face, name, and personality of a new colleague at work. You might think I'm waxing a little metaphysical here, but I believe that daily, monthly, and yearly cognitive challenges are no more or less valuable than emotional experiences in the same temporal degrees.

There are 36,525 days in a hundred-year life. As it happens, that's not so far from the 34,000 distinct emotional states that some psychologists have suggested humans will experience over a lifetime. We should each make an effort to feel them all.

EXERCISING EMOTIONAL PLASTICITY

Eric Kandel was just ten years old in 1939 when his family emigrated from Austria to the United States as Nazi Germany increased its militaristic footprint and consolidated its hold on much of Europe, placing Jewish families like the Kandels at the gravest of risk. As often happened in such times, one nation's loss was another's gain: Kandel would go on to become one of the world's most influential brain researchers, a recipient of the 2000 Nobel Prize for his work on how neurons store memory. Kandel returned to Europe many times during the war, but perhaps no trip back to his homeland was so profound as the one he took in early 2024 when he traveled to Vienna to receive the Decoration of Honour for Services to the Republic of Austria, among that nation's highest awards for civilian service. He was ninety-four years old and, by all accounts, as spritely and sharp as ever.

I reflect here on Kandel's remarkable life not simply because he is yet another cognitive guide star—who actively continued his research well past the age that most have retired—but because even into his nineties he was still trying to convince his academic peers to correct what he saw as a grievous mistake of medical history, the century-old divide between neurology and psychiatry.

Kandel has long argued that these disciplines should be viewed as one and the same, or at most they should be seen as subspecialties of the same tightly connected field. "All mental processes, even the most complex psychological processes, derive from operations of the brain," he wrote in 1998 and reiterated many times thereafter. He was right then.

He is right now. Training our brains to remain usefully and youthfully plastic requires us to work just as much on our mental health as our neurological health.

Neurology and psychiatry were, in fact, once the same specialty, but split just over a hundred years ago. As we understand more and more about the underlying drivers of neurological and psychiatric diseases, it is becoming clear that these two should once again be united.

Early in my career, I assumed that the patients I saw were so commonly depressed because they were dealing with the revelation that their lives were coming to a brutal sort of end, a gradual or not-so-gradual loss of the very substance of who they were. Of course they were depressed. Who wouldn't be depressed? It later occurred to me that it was fully possible that these unfortunate souls were depressed long before they began losing their functional cognitive health. This has since been well established by researchers, including Stanford University epidemiologist Victor W. Henderson, whose work has demonstrated that the risk of dementia more than doubles for individuals who have a previous diagnosis of depression, often many decades before a neurodegenerative diagnosis.[25] This is probably only the tip of the iceberg, as it does not account for those with undiagnosed and thus untreated mental illness, a group that is likely even far more at risk of long-term diseases of the brain and body.

We have also come to know, however, that the biochemical changes that lead to dementia often happen many years before those symptoms occur. And so, it would at first seem as though we have something of a chicken-and-egg riddle on our hands: Does depression lead to neurodegeneration or does neurodegeneration lead to depression? Some might observe that it could be both—it is not at all unusual for two conditions to be independently precursive of each other. But there is another intriguing possibility: Perhaps these two conditions share the same root disease.

This theory aligns with the way that an increasing number of researchers are coming to understand many chronic diseases, and especially those illnesses that are often considered diseases of aging. Not too far upstream, this line of logic goes, it is not cancer or heart disease or osteoarthritis or macular degeneration—it is aging itself. Likewise, I have found it increasingly difficult to conceptually separate psychological ailments and neurological diseases. Not too far upstream, after all, these conditions look very much the same—they are biochemical dysregulations that come as a result of a network of insufficiencies, leading to a cascade of failures that make life harder and harder to live.

If depression is an initiating cause of neurodegeneration, then treatment for depression should be in some ways preventative. That would seem to be a correct inference. But this does not preclude the possibility—indeed, as I see it, the likelihood—that treatment for depression, in particular psychological therapy, mitigates both the immediate symptoms of depression as well as the underlying cause of neurodegeneration.

If that sounds like neuroplasticity to you, I think you're right. And we're not alone. A growing body of evidence connects depression to the atrophy of neurons in the cortical and limbic regions of the brain—those that control mood and emotion. Effective therapy, meanwhile, has been shown to reverse the neuroanatomical indicators of this loss of neuroplasticity.

I hasten to emphasize the word *effective* from the above paragraph. Owing in no small part to the unfortunate divide between neurology and psychiatry, the latter is woefully uneven in approach and evidence. I personally know people whose lives were unquestionably saved when they found the right therapist. I also know individuals who feel their lives were harmed, in some cases irreparably, by a lack of effective psychological care, and these are sadly not unusual cases. Writing in the journal *Frontiers in Psychology* in 2019, one team of researchers recounted patient experiences defined by feelings that they were not in the right kind of therapy, that their goals were not being met, that they were not being listened to, and that even when there was no actual malpractice, their experiences left them feeling much worse off than they were when they began the process. As one service user said, "It's like a lottery, only you can either gain big, lose big, or land anywhere in between."

Experiences like these left some patients questioning what, if anything, was guiding psychological treatment. "We've all been told that this baloney somehow is on the same par with medical services," that patient said. "They've been trained and validated by prestigious institutions. Much of what we watch and read tells us these are serious, qualified, responsible people who will improve our lives if we follow their program." That patient had clearly concluded that this was not at all the case.[26]

To be clear, these are experiences that are replicated in the lives of those who receive other kinds of medical care. The difference is that most medical experiences can also be quantified to at least some degree by biochemical data. Whether or not your oncologist had a terrible bedside manner, if the care led to remission of your cancer, it was successful in at least that respect. A robust integration of biodata reflective of a rested and restorative mind—from stress hormone levels to sleep cycle monitoring—would be a good place to start for those who are rightfully interested in knowing whether, at a physiological level, their therapy is working to mitigate psychological distress or disease.

To this end, I am quite confident that any sort of brain training that is sufficient to prevent cognitive decline is necessarily both neurological *and* psychological. This is to say, such supportive activities not only help improve memory, attention, and learning but also mood, social connection, and mindset. A well-trained brain—one that is structurally resilient—is one that is psychologically resilient.

FRUSTRATION

I thought I might lose Douglas, the former contractor I described at the onset of this chapter, on the day that his son, David, expressed his frustrations with the lack of rapid improvement in his father. But David agreed to keep his father on the protocol. For a period after we had begun treatment in 2012, I recommended that patients who had progressed beyond SCI and beyond MCI to develop dementia not be treated, since interventions are more difficult, and the results are often less dramatic. However, after receiving a few emails from clients' family members, all of whom noted how important simple changes were—more family engagement, return of speech, return of continence, return of cooking or dressing—I reconsidered this approach.

I remember one email in particular, because the husband was quite critical. In his message, he told me that I should never suggest limiting treatment to those with SCI or MCI, because his wife, with a MoCA score of 0, had reengaged with the family after adopting the protocol. I took that to heart.

I have many stories of patients who experienced substantial cognitive decline only to rebound to a more youthful cognitive state as they embraced the ReCODE principles. I am sad to say that Douglas never made a complete recovery, but the difference we were able to make in his life was substantial. He was able to resume many of the joys and experiences of daily life, especially his ability to carry on a conversation with his children and grandchildren. In light of these successes, which did indeed eventually permit us to engage Douglas in some more formal brain training, David seemed to soften toward me.

I am always a little bit haunted by what might have been with patients who do not make recoveries. I do wonder what more we could have done. In Douglas's case, I have a few ideas about the power of exercising neuroplasticity in ways that are not so taxing on an overwrought brain that I wish we could go back in time and employ.

I especially remember the ones we have lost. I remember Barbara. And Becky. And Richard. And Uta. And others. What did we miss that might have made the difference? What could we have done for better outcomes? We are still working on those questions, every day.

There's an emotion for this—something related to regret but colored by wonder—that I am still getting to know. I'll be sitting with it for a while.

11

THE TOXIC ADVENTURE

> Education, whatever else it should or should not be, must be an inoculation against the poisons of life.
>
> —HAVELOCK ELLIS

One of the developments that has been most surprising to me in our studies of neurodegeneration is how common toxins are as contributors to cognitive decline. From air pollution to mercury, anesthetic agents to toluene, trichothecenes, and hundreds more, these toxins are vastly understudied and undertreated as contributors to brain aging and cognitive decline. As such, chemicals often turn out to be contributing factors in the messages I receive, multiple times each week, from people who have had a cognitive "wake-up call." This could be a series of important meetings that they've missed, a loved one who has worriedly confronted them about their inattention, or something equally jarring. Often, these individuals have already been piecing together information they have found about the ReCODE Protocol online, but they soon discover that they need more help.

The internet is a Wild West of information, but there are, in fact, many resources out there about our and others' work that are accu-

The Toxic Adventure

rate, informative, and useful. One such source is the site ApoE4.info, founded by Julie G., a friend, citizen scientist, and patient who has sustained remarkable cognitive improvement for over twelve years. Julie has also created more than one hundred guides for everything from sleep to detox to supplements, which are available at Apollohealthco.com.

Because there is so much information out there, it's not unusual for people to have experienced some early successes before they begin working with a trained practitioner. For some, making dietary changes is enough to help improve memory, attention, and clarity. For others, diet plus exercise can be a game changer. For others it takes changes to diet, exercise, and sleep. But for many, unfortunately, those steps aren't enough because—as you know—Alzheimer's is a systems-wide failure that comes as the result of myriad cognitive insults. By the time it's noticeable (and we finally ignore the whispering demons of denial), the structural elements that make a person's brain work as it should have been degrading for decades. Rebuilding those structures takes time and effort. The seven basics—a plant-rich, mildly ketogenic diet; exercise; optimal sleep; destressing; brain training; detox; and targeted supplements—often lead to temporary improvement, but long-term recovery usually requires addressing additional, specific underlying contributors as well.

One commonly overlooked contributors to neurodegeneration is toxin exposure, because the environments in which we live and work are often places where we are exposed to contaminants. It's usually much harder to change your environment than it is to change your diet, exercise, and sleep habits. So, when we assess the bloodwork, conduct a liver panel, study urinary toxins, do a home inspection, and take other measures to evaluate exposures for people who haven't responded to the basic metabolic optimization, we often see that individuals have very high rates of dangerous metals, volatile organic chemicals, and biotoxins in their bodies.

That was the case for Karl, a financial analyst in his fifties who had been appearing as an expert on cable television news stations before he began losing focus and clarity. Improvements to his diet, exercise, and sleep had stabilized Karl's situation. Subjectively, it seemed to him and his loved ones that he was actually getting a little bit better, but not to the point that he could fully reengage in his life's work. Something was still missing, and when he began working with a ReCODE-trained health coach in New York, it soon became clear what that something likely was.

Karl's blood tests revealed a high level of matrix metalloproteinase-9 (MMP-9), an enzyme known to be associated with arthritis, coronary artery disease, chronic obstructive pulmonary disease, multiple sclerosis, asthma, and cancer. It has also been implicated as a marker of inflammation in people with diseases that appear to have been triggered by biotoxin exposure, along with transforming growth factor beta 1 (TGF beta 1) and a fragment of the Complement component 4 protein (C4a).[1]

Biotoxins are chemicals produced by plants, bacteria, and fungi as protection from predation and niche competition. Nicotine, for instance, is a biotoxin that is produced in the leaves of nightshades, most notably tobacco, while tetanus is a biotoxin carried by the bacterium *Clostridium tetani*. But the biotoxin most associated with cognitive decline is mold—not all molds, but several types of mold, including *Stachybotrys, Aspergillus, Penicillium, Chaetomium,* and *Wallemia* that produce mycotoxins. They appear to wreak particular havoc on the human brain. In 2016, having come to recognize that many of the patients I was seeing had been exposed to these mycotoxins, I wrote a paper that sought to bring this common cognitive insult to the attention of the wider medical community.[2] When "Inhalational Alzheimer's: An Unrecognized—and Treatable—Epidemic" was published in the journal *Aging*, there was quite a bit of skepticism. But since that time, we have seen that mycotoxins contribute to cognitive decline in nearly half of all cases, and it's becoming far less common to hear physicians dismiss this insult out of hand.

What in evolution would cause some molds to produce toxins that harm humans? Well, if you are a mold cell, you reproduce more slowly than the bacteria around you, so you are going to lose your real estate fairly quickly unless you produce something that kills or keeps the bacteria at bay. This is why the world's most famous antibacterial substance, penicillin, is produced by mold. That's good for mold, but it's bad for us, because the mitochondria in our cells are descendants of bacteria,[3] and thus at increased risk of being damaged by these mycotoxins. And with compromise of our mitochondria comes reduced energetic support, which is a key driver of cognitive decline.

If we are to prevent those declines altogether, ensuring long and healthy brainspans, we must identify the toxins, remove them from our bodies, and do everything we can to eliminate them from the places where we live and work. But while I have been excited to see a swell of doctors in recent years take up the fight against mycotoxins (not just to prevent cognitive decline but other ailments as well) this common contributor of disease is often still ignored by the so-called standard of care.

In fairness, the connection between mold and disease is a lot more complicated than it might first appear. Some fungal species can produce multiple different mycotoxins, especially when the environment in which that species lives changes, and many different species are capable of producing similar toxins,[4] so there is widespread disagreement on how to even organize the species, the toxins they create, and the symptoms thereof.[5] This lack of a neat-and-tidy classification renders cause-and-effect diagnoses quite difficult, and so after decades of debate, physicians are still arguing over whether mold is really that big a problem at all. Alas, the whitecoats will probably be debating this for decades to come.

In the meantime, let me tell you what I have seen again and again: Mycotoxins are in the bodies of people with cognitive decline, and the mold toxins are in their homes and places of work. And these observations align with a striking study, published in the journal *Scientific Reports* in 2015, revealing that tissues from the brains of people who died with

Alzheimer's disease often contain fungal cells that are rarely detected in controls.[6] What's more, I've seen that reducing these toxins often leads to cognitive improvement, and I'm not alone in that observation. Jill Carnahan has seen it. Mary Kay Ross has, too. Mary Ackerley, Kat Toups, and Craig Tanio are just a few of the other practitioners who have publicly cried out for more attention to this pervasive insult. The movement is growing so quickly, in fact, that there is an entire society that is largely devoted to mycotoxin illness, the International Society for Environmentally Acquired Illness (ISEAI).

Indeed, not long after learning of the disconcertingly high levels of MMP-9 in Karl's bloodwork and setting to work searching for the potential environmental contaminant that might have been the cause, I received an email affirming that the ducts in Karl's home were indeed overrun with mold. It's too early to know whether mitigation of this problem will result in further and more substantial improvements in Karl's case, but I'm feeling optimistic.

I suppose that Karl is both unlucky and lucky. He was, it would appear, quite unfortunate to have been living in a home that was infested with fungal species, even more because—for reasons we don't yet completely understand—some people are more susceptible to mycotoxin exposure than others, likely because of their immune responses, as coded by the presence of specific sequences in their DNA. (These associations have been shown in some types of cancer,[7] and, given that we often see people in the same home with vastly different reactions to spore exposures, it stands to reason that some people are genetically inclined toward cognitive insult as well, but the research in that and other areas of human health remained very limited.) Karl is fortunate, however, to have had the means to relocate while his home was professionally eradicated of that source of contamination and HEPA filters were installed to minimize future exposure to mycotoxins, molds, mold spores, and other irritants that cause inflammation—which are collectively called inflammagens. If the problem persists—because it can be difficult and in some

The Toxic Adventure

cases impossible to eliminate mold once it's taken root in a building—he may need to move to an entirely new home.

Few people enjoy that level of privilege. It's not easy for most people to move homes and jobs. It's even harder to move away from a city or region where the air, water, or soil is heavily polluted. (These are the most common vectors for the metals and volatile organic chemicals that we will also discuss in this chapter.) What's more, many of us—if not all of us—have been exposed to scores of pollutants for our entire lives, including before we were even born.[8] All of this is why it is so very important to do whatever we can to reduce our toxic exposures from the earliest points in our lives. Many toxins are, after all, something we will carry with us for years and decades after exposure in our blood, our bones, and our brains.

And so, when it comes to this particular cognitive insult, every day counts.

MITIGATING MYCOTOXINS

It wasn't long after Katie moved into enlisted family housing at an Air Force base in northern Utah that her two daughters began repeatedly falling ill, both to the point of requiring frequent visits to the base hospital. Having heard that other families on the base were suffering from health conditions that seemed to be related to mold exposure—by far the most common form of biotoxic exposure—Katie suspected she'd identified a potential cause of her daughters' illnesses, but officials from the private company that runs housing on the base declined to even examine her quarters. So Katie hired a private inspector and, sure enough, the results showed disconcertingly high levels of toxic spores inside her home.

The housing officials weren't moved, and even accused Katie and others with similar complaints of trying to use inspection reports and health records to upgrade to a nicer, newer home. Katie was livid at that contention—she had decided to break her lease, at significant cost to her family, to move *off* the base. "We can't wait to get out," she said in 2007.[9]

Over the years, stories like Katie's accumulated at military bases across the United States, and the issue eventually came to the attention of Senators Elizabeth Warren and Thom Tillis, who teamed up to hold private providers accountable for poorly managed military housing. Perhaps not surprisingly, it turned out that the private company that showed no interest in inspecting Katie's home was one of five contractors implicated in violations of a "tenant bill of rights" for military families in a bipartisan-supported investigation.[10]

Even if Katie's actions hadn't helped begin the process of revealing these sorts of issues, she still made the right choice to leave the base housing. The challenges of connecting specific symptoms to specific mycotoxins is complex, but it has long been known that exposure to mold spores can trigger difficulties with breathing and other allergic reactions in many people. Additionally, we now know that chronic exposure to biotoxins produced by molds (as well as viruses and bacteria), are linked to neurodegenerative diseases. In fact, one of the biotoxin-causing molds found in Katie's home, *Aspergillus*, has been implicated in a litany of cognitive insults, including oxidative damage, inflammation, and higher rates of cellular death.[11]

Maja Peraica, the head toxicologist at the Institute for Medical Research and Occupational Health in Croatia, which has seen a frightening increase in foodborne mycotoxins over the past fifteen years,[12] has pointed out that children are likely much more vulnerable to biotoxins because they have lower body mass and have not yet developed a mature immune system.[13] That may help explain why a cohort study of children in Poland showed that long-term exposure to mold in infancy and during the toddler years was associated with a tripling of risk for low cognitive functioning at the age of six, even after adjusting for factors that are known to be correlated with cognitive development, including maternal education, breastfeeding practices, and exposure to tobacco smoke.[14] Another common mycotoxin, ochratoxin A, has been implicated in increased rates of autism in children.[15] But while children are particularly susceptible, even exposures that happen in adulthood can constitute a major insult for many people.

It might be useful at this point to turn back to our analogy of the uphill bike ride. As you'll recall, the basic idea is that cumulative insults make it harder to bear the weight of the cognitive load we're carrying up the hill. Eventually, something has to be left at the side of the road. So far, though, we've presumed that we always start at the bottom of the hill with a certain amount of energy. If, at the point that the road begins to become more steep, you've *already* been riding for a long time, carrying a large load while hampered by a flat tire, making it up that hill becomes more precarious.

When it comes to mold, what can we do to make the ride a little easier? Well, the obvious first step is cleaning. The longer mold is permitted to persist, the more opportunity it has to spread, so all visible mold should be addressed immediately—scrubbed from hard surfaces with detergent and water while wearing an N-95 respirator mask and providing as much ventilation as possible. For softer and absorbent surfaces, there is seldom a sufficient solution other than to throw away that which has become moldy. If the mold source is in the plumbing, any leaks should be fixed and the mold-damaged area around it cleaned or removed. For anyone having their home remediated, it is crucial to be out of the home during the remediation, because transitory increase in mycotoxins during remediation can make cognition much worse.

This is a battle many people will have to fight again and again. As is often pointed out, we live in homes made of "mold food." Fungi love wood, drywall, wallpaper and carpets, and pretty much any space with a little bit of moisture, a lack of ventilation, and darkness. I'm not convinced that most people are ever in a position to completely eradicate the most common sources of biotoxic exposure. We can, however, be more thoughtful about the places in our home where we choose to spend the majority of our time. And there is one area of the home that is typically the most prone to mold: its subterranean layer.

Over recent decades, there's been an increase in young adults deciding to move back to their parents' homes, and there's a very good reason why "moving into the parents' basement" has become the operative cliché for this phenomenon. I've also noticed with some dread that people seem to be increasingly eyeballing their basement as a good place for conversion into rental apartments or as a place where older family members can "age in place" close to relatives. And another recent trend is the advent of the über-chic basement home theater—the fanciest of which include curtains, lush carpeting, sound panels, drop ceiling tiles, and super-plush recliners, all of which turns the moldiest place in most homes into an even more inviting space for spore making, especially once someone spills a beverage during a horror movie jump scare. As the COVID-19 pandemic pushed tens of millions of people to work from home—and many decided they liked that arrangement—basements have also become one of the most popular places for a home office. I fully understand the social, familial, and economic forces at play in these decisions. I know that square footage is at a premium. However, if you have decided to utilize basement space in these sorts of ways, I'd advise using an inexpensive do-it-yourself mold testing kit on a quarterly basis, and either an Environmental Relative Moldiness Index (ERMI) testing kit or a Health Effects Roster of Type-Specific Formers of Mycotoxins and Inflammagens (HERTSMI-2) kit every few years or any time in which nasal and sinus congestion, difficulty breathing, skin irritation, or headaches prevail for longer than the span of a common cold. If you haven't already used a basement in this way, I suggest factoring the potential for mold—and the possibly associated cognitive consequences—into your decision-making. Basements are great for furnaces, tools, long-term storage, and keepsakes you don't use but can't stomach throwing away. These are not ideal spaces for spending hours and hours of our lives.

If mold is an issue in your home, the next step—a much harder one, to be certain—is to consider whether a move is possible. One of the things that has always baffled me is that almost everyone, once they have the means to do so, will undertake the stressful experience of moving, even if it's within the same general area where they currently reside, when they feel they need a bigger home or want a nicer home. Yet very few people will embark on that adventure explicitly to protect their health. If mold is a persistent problem in your home, cognitive health problems are *much* more likely. If at all possible, don't wait to escape!

MANAGING METALS

Now that we've broached the subject of moving, it's probably a good time to bring up some of the other toxins that anyone who wishes to have an ageless brain should aim to avoid. If it is at all possible to do so, consider leaving highly polluted homes, neighborhoods, and cities.

This isn't always necessary, let alone possible, and in most cases we really can manage and mitigate our exposures in place. But there are extreme cases in which people have understandably decided to just pick up and leave.

That's what thousands of people from Flint, Michigan, did in the years after that city switched its water source to the Flint River. This shift to water that was different in mineral and chemical composition did a number on the old pipes of the distribution system, causing municipal water to become contaminated with lead, a heavy metal that is well known for causing brain damage. For many months after the change, city and state officials denied there was a problem, but accumulating evidence from residents who took it upon themselves to test the water—and later confirmation from academic researchers who did system-wide tests of the water coming from people's taps—affirmed that the city had effectively been subjected to widespread poisoning. State Senator Jim Ananich, who represented Flint in the Michigan Legislature, wasn't surprised when residents began to flee. "What we've gone through, especially with the water crisis with folks being lied to and mistreated, I cannot say I blame people for considering their options," he said in 2021.[16]

I don't blame them, either. I almost certainly would have done the same thing, and my heart breaks for those who didn't have the economic means to leave if they felt it was in their family's best interest to do so. I have seen the consequences of heavy metal exposure on the human brain.

When twenty-one of us, as coauthors, published an accounting of one hundred patients who had experienced a reversal of cognitive decline in 2018, many of those whose stories we shared had first tested positive for high levels of heavy metals in their bloodstreams.[17] That was hardly a surprise, since in epidemiologic studies going back for decades, high exposures to metals like lead, cadmium, manganese, and mercury have been demonstrated to be associated with cognitive decline. There are many reasons why these metals are so toxic to the human brain. They are known to "gunk up" the signaling networks within and between cells. They increase oxidative stress. They disrupt the essential function of the metals that are supposed to be in our bodies (or, in the case of manganese, which is an essential mineral, at much lower levels). They trigger brain inflammation. They cause early cell death. They disrupt the epigenetic directors of genetic activity and cellular identity.[18] While we could certainly benefit from more longitudinal studies demonstrating the cognitive consequences of the effects of exposure, there's really no serious debate on this point. Heavy metal exposure is a major insult that chips away at cognitive health.

And it is one that starts at our tap. If the home or building you live in was built before the mid-1980s and you're not sure it's been completely overhauled with new plumbing, you may still have some lead pipes carrying your drinking water. But it's not just people in old homes who are at risk; many people in Flint were shocked to learn that their city's water distribution system was still replete with lead pipes, but the reality is that the east-central Michigan city is not unusual. Decades after US environmental regulators banned new lead pipes from being installed, the replacement of the old ones is still underway. Marc Edwards, the Virginia Tech environmental and water resources professor who helped expose the Flint crisis, has pointed out that no one actually knows where the old lead service lines are, because federal law never required a thorough inventory.[19] And even if the pipes aren't the problem, heavy metals can be present in the source of the water, and aren't always adequately captured in water treatment facilities.

I strongly recommend using a heavy metals water testing kit, which can be purchased for less than $20, about once per year to understand the potential metals in your water. If there is any indication that your water is high in any heavy metals, filter your water. And, for peace of mind, you might want to filter anyway, especially because tests are nothing more than snapshots of water quality at a particular time on a particular day, and these levels can fluctuate greatly in different sources.[20]

Of course, water is not the only potential source of heavy metals. Cadmium, lead, and mercury are some of the most common air pollutants that are emitted from vehicles, factories, and energy plants around the world. In many cases, these and other metals that have leached into the soil can become airborne when disturbed by vehicles or heavy winds (the latter of which is especially worrisome, as some research has suggested surface winds may be increasing as global temperatures rise.[21]) While cancer and other chronic diseases are more commonly associated with exposure to these airborne particulates, in recent years increasing attention has been paid to the cognitive consequences. In 2021, for instance, an international team of researchers warned that people living in metropolitan Mexico City—which was once widely regarded as the world's most polluted place, with high levels of metal-contaminated air—were suffering from much higher rates of cognitive impairment than should be expected of people living in similar but less-polluted environments. Researchers are typically reticent to engage in hyperbole, and a review of the other peer-reviewed papers published by members of this particular team reveals no predilection for histrionic rhetoric, but in this case the team was adamant. "Our results are disturbing," they wrote. "The issue should be an urgent public health research priority and the impact on health, educational, social, economy, and the judicial systems ought to be a serious concern."[22]

I absolutely agree. While it is reasonable, and I think largely correct, to suggest that individuals have a responsibility to take care of themselves when it comes to many lifestyle choices, there is less choice in the air we breathe. And for many people living in nations where air pollution is rampant, moving is not an option. The best one can do is wear a well-designed mask during pollution spikes or all the time in places where pollution is rampant. It's true that masks can be effective against common pollutants,[23] but this is not a palatable or scalable strategy in the long term. Truly, the only solution is a collective one, and as air is a globally shared resource, it absolutely must be a global solution.

Most of us are in a much better position to care for ourselves when it comes to fish, one of the most prevalent sources of heavy metals, and particularly mercury. I should note here that I love seafood, and one of the great joys of living in a coastal state is having regular access to fresh fish and shellfish. There's little doubt that many of these fish—particularly those that are high in omega-3 fatty acids such as salmon, mackerel, anchovies, sardines, and herring—are beneficial for cognitive longevity.[24] But it has also long been known that on-shore industrial activity that spews mercury into the atmosphere has an outsized effect on our oceans, where plants and animals at every trophic level are literally swimming in a sea laced with this heavy metal, which is the only element known to be liquid at standard temperatures and thus is easily circulated in the ocean. As we discussed in chapter 7, as small animals are consumed by larger ones, and larger ones are consumed by even larger ones, this toxic metal bioaccumulates in living tissues and the result is that predators toward the top of the food chains carry *a lot* of mercury in their bodies. And that means they can deliver it to our bodies.

Mercury exposure can bring about the neuropathology of Alzheimer's, and our exposure can be to organic mercury—from seafood sources like tuna, shark, and swordfish (long-lived, large-mouthed fish)—or from inorganic mercury from dental amalgams. It's worth noting that APP, the molecule at the center of Alzheimer's disease, which gives rise to the amyloid we associate with Alzheimer's, is actually a metal-binding protein that interacts with iron, copper, and zinc, among other metals. Thus, there is an intimate relationship between metal exposure and Alzheimer's-related brain signaling.

Like most cognitive insults, a diet high in mercury isn't likely to be the *only* thing that results in a brain that ages faster and degenerates more—and won't affect everyone in the same way—but it is a nonetheless likely cofactor for many people, including those with a common gene variation, known as Val66Met, in the gene that codes for the production of brain-derived neurotrophic factor, which plays a major role in the maintenance and survival of neurons.[25] Unfortunately, we don't yet have enough population-scale genetic testing to be able to know with greater certainty who might be most at risk, which is why I advise caution with consuming fish like wild tuna, marlin, shark, swordfish, sturgeon, and halibut. For most people, it would appear, occasional helpings of these fish are unlikely to be an issue, but it's worth remembering that when we eat these animals, *we* are becoming the apex bioaccumulator.

MICROPLASTICS ARE MACROPROBLEMS

Toxic metals aren't the only things that bioaccumulate in our environment and, ultimately, our bodies and brains. Another growing concern is microplastics, fragments of plastic that have broken down into smaller and smaller pieces, often to the point that they become nanoplastics, or microscopic fragments. At these scales, microplastics easily find their ways into seafood, processed food, bottled water, and many other things. In one study, nearly 90 percent of all protein sources sampled, including plant-based meat alternatives, were found to contain microplastics, and the study's authors estimated that American adults could be consuming up to 3.8 million pieces of microplastic per year from these protein sources alone.[26] That would be reason for concern even if those tiny plastic pieces were moving into and out of our bodies through our digestive systems, but that's not what's happening. They're circulating in our bodies. One study found that people with vascular plaque were far more likely to have an adverse cardiovascular event if their plaques contained microplastics.[27] They're also in our lungs. Another study detected microplastics embedded deep in human lung tissue,[28] where it may make breathing harder, straining the supply of energetics brains need for youthful functioning. Studies on mice suggest these tiny pieces are also easily penetrating the blood-brain barrier; after just four weeks of drinking water contaminated with microplastics proportional to what many humans ingest each week was enough to result in microplastics getting lodged in the animals' brains.[29] A team of researchers led by Jaime Ross, a neuroscientist who studies the causes of brain aging at the University of Rhode Island, reported in 2023 that particles like these induce cognitive shifts and physiological changes in the brains of mice, including a decrease in glial fibrillary acidic protein (GFAP), the protein I first mentioned in chapter 6 that is so useful as a marker of inflammation in the brain.[30]

Do we know everything about microplastics and human cognition? Not even close. But do we know enough that we should be doing everything we can to avoid this toxin? Without a doubt.

So, how do we accomplish that? Well, it does appear that many of these plastics are entering our bodies from our food, and the processing and packaging that happens to that food between farm and table, so a focus on organic foods and fresh foods, and avoidance of plastic-packaged and processed foods, is wise. (I also suggest avoiding plastic containers, straws, and bags. Never microwave food in plastic and use only glass or metal for food and drinks.) Increasing your intake of fiber both soluble and insoluble—at least 30 grams per day—may also help, as foods such as leafy greens, legumes, nuts and seeds, and supplements like organic psyllium husk or konjac root, can help keep microplastics from escaping the gastrointestinal tract and making their way into your bloodstream. Cruciferous vegetables such as brussels sprouts and cauliflower are additionally known to offer detoxifying benefits.[31]

It's not yet clear whether there is a most effective route for getting rid of all the plastic we accumulate, but it's important to note that the plastic itself isn't the only problem—it's also the chemicals that come along with it, like bisphenol-A (BPA), a compound that has been shown to affect memory and cognition and alter levels of vital neurotransmitters.[32] But BPA can be expelled from our bodies, and sweating may be the best mechanism for doing so (but beware, some manufacturers have simply substituted other, similarly toxic chemicals for the BPA).[33] Deep breathing may also be effective to reduce pulmonary microplastics—I recommend breathing in through your nose for four seconds, holding for four seconds, and then exhaling slowly over four to eight seconds, a practice that is valuable for other aspects of mental and physical health as well. And, of course, there's no better way to amp up your respiratory rate than exercise.

It's true that microplastics are so pervasive that it's virtually impossible to avoid ingesting them. They are, alas, a form of pollution that we may never fully mitigate. But it's one thing to acknowledge that we can't completely avoid a problem and it's another to throw your hands up and say, "There's just nothing that can be done." That's not true. We can do a lot to avoid microplastics and expel them and the chemicals they carry from our bodies—and to assure cognitive protection and performance for life. We absolutely must engage in this fight.

VANQUISHING VOLATILE ORGANIC COMPOUNDS

Almost all cases of cognitive decline are identified in one of two ways: Either the person stops resisting the accumulating evidence of memory loss, difficulty planning, word-finding problems, problems with facial recognition or navigation, and seeks help for themselves, or someone close to them identifies these or related symptoms and seeks help for them.

That's it. Few doctors talk to their patients about cognitive decline in a proactive way, even though the vast majority of people say they would be receptive to these important conversations,[34] and even fewer physicians actively suggest cognitive testing for their patients. There is no widespread screening underway—not in any nation in the world! This must change if we are to reduce the global burden of dementia.

But when the US Preventive Services Task Force examined the question of whether there should be testing of this sort, its answer was a very tepid "maybe."[35] This, to me, is rather flabbergasting, for it means that we learn of people's conditions only when the insults have accumulated. However, I do concede that comprehensive testing of *everyone* is, for now, a resource-intensive nonstarter. Even a ten-minute MoCA test takes longer to administer than some physicians spend in total with their patient in a single visit,[36] and reading and interpreting a MoCA takes additional time.

Perhaps one day we'll realize our folly. For now, a research team from Xiangya Hospital in Changsha, China, approached the question in a very practical way. In a country of 1.4 billion people—where the number of people over the age of sixty is greater than the population of all but three other nations in the world—what might be the least expensive way to test as many people as possible?

One potential answer, they suspected, might be to conduct a very inexpensive breath test for volatile organic compounds (VOCs), chemicals that are usually man-made and that have a high vapor pressure and low water solubility, and thus easily circulate in our environments and just as easily find their ways into our bodies. There are thousands of VOCs, but some of the most common are benzene (which is used to make plastics), ethylene glycol (an essential ingredient in polyester and many antifreezes), formaldehyde (used for everything from embalming bodies to dyeing textiles to strengthening wood products), and methylene chloride (for all your paint stripping, metal degreasing, and pharmaceutical manufacturing needs). VOCs have been implicated in nasal tumors, leukemia, asthma, nasopharyngeal cancer, and reduced pulmonary function,[37] but so far the research on the connections between these chemicals and cognitive aging is limited, with much of the research simply noting VOCs as part of a panoply of airborne insults that may contribute to cognitive decline.[38] And yet when the Chinese research team took breath samples from 1,467 people, they discovered that those with a previous diagnosis of cognitive decline were exceedingly likely to also have a high concentration of ten VOCs in their breath tests—results affirmed through a neurofilament light chain test, one of the tests I mentioned in chapter 6.[39] That result points to the dire need for further investigation of the mechanisms by which VOCs imperil cognitive health, but it also points to the fact that this inexpensive test can and should be used now to assess neurodegeneration in a preventative way.

Testing for VOCs, either with a well-designed meter or through a professional contractor, is a step I recommend for everyone at least once every few years. Fortunately, if your home's VOC levels are low, it's pretty easy to keep them that way—by refraining from smoking, using natural and nontoxic cleaning products, storing chemicals and pesticides away from your home, and only bringing in new furniture, decor, and clothing from toxin-free brands. It is also helpful for all of us to have HEPA-type filters, ones with cartridges that remove both particulates and VOCs. If you do have detectable levels of VOCs, mitigation is often as simple as keeping your home well-ventilated, replacing furniture that continues to off-gas VOCs long after it has been purchased, and sealing off walls that have previously been painted with non-VOC paints.

TOSSING TOXINS

We live in a toxic world. Long before our species became such a dominant force on this planet, there were plants, animals, fungi, and microorganisms already here that were willing, able, and ready to protect their own survival through the production of chemicals that were poisonous to other living things. There were heavy metals in the Earth and volatile chemicals in the air. How we managed to survive all that is a wonder, but we did it.

Then, without much understanding of what we were doing, we began *adding* ingredients to this toxic soup. Even after we did understand, we kept right on going, pulling even more metals from the ground, putting even more chemicals into our air, spreading plastics everywhere, and producing even more volatile compounds to bring into our homes in a two-steps-forward-who-knows-how-many-back effort to make our lives a little bit better.

This isn't just a dance we do with the toxins that we are accidentally consuming and inhaling; sometimes we actively and intentionally ingest things that harm us, from alcohol and cigarettes to illicit drugs and prescription medications—even those taken specifically as directed. What is a pharmaceutical that has more detrimental effects than beneficial ones? It is nothing more than another toxin. And what is a pharmaceutical that is initially more beneficial than detrimental, but that accumulates in its original form or as a metabolite in our livers, kidneys, blood, and brains over time, causing other systems to fail? It is also nothing more than another toxin!

Again, I want to reiterate that I am not *at all* against medication—indeed, I believe the future is to combine personalized, precision-medicine protocols with targeted pharmaceuticals. I am, however, quite vehemently opposed to medications that do not work very well, if at all, in comparison with the toxicity they introduce into bodies and brains that are already awash with toxins. When people understand pharmaceuticals in these ways—as sources of potential toxicity alongside potential benefits—they are better able to share the decision-making process with their doctors, and more careful about choices that can result in prescription drug–induced cognitive impairment.[40] As part of that process, it's always appropriate to ask your doctor about the known and suspected cognitive consequences of a drug, and whether there is a plan for ultimately getting you off of that medication once you're on it.

The health effects of the toxic adventure that our species is living through have been piling up for a long time. But it's pointless to complain about any of this. What is done is done. The question is: What can we do now?

Thankfully, as individuals, there's a lot we can do. We can begin the process of eliminating the main sources of toxins from our homes and places of work. When and if we move, we can be intentional about our choices. We can measure our environments, from time to time, to make sure we're doing a good job of keeping heavy metals, VOCs, and biotoxins at bay, and stay abreast of research that reveals new risks, like microplastics, and new ways to mitigate those dangers. We can eat healthy foods, exercise often, get plenty of restorative sleep, and eliminate sources of negative stress in our lives, thus providing our bodies with the resources they need to detoxify. While not yet well studied as a means of detoxification, many people report that they use saunas for that very purpose,[41] and given the quite well-known cognitive benefits of sauna bathing,[42] it stands to reason that this method of "sweating it out" might indeed be playing a detoxifying role.

If and when all that is not enough, we can work with physicians who understand and embrace our detoxifying goals to accentuate the pathway to healthier bodies and ageless brains. You may wish to discuss with your physician strategies for the increased production or supplementation of glutathione, which may help transform toxins into mercapturic acid, which is more easily flushed from the body. It's also important to know that different toxins stay in our bodies for different amounts of time, ergo there may be times in our lives where we will need different detoxification strategies. One such example is menopause, which comes for most women along with significantly increased rate of bone degradation. Unfortunately, that's where the human body stores a lot of the heavy metals it has absorbed (which is why we can trace historical changes in lead production by examining human remains from thousands of years in the past[43]) and this "osteoclastic burst," in which toxins are being rereleased into the body, may help explain why we often see a swift shift in cognitive well-being during the menopause transition, particularly when it comes to psychomotor speed,[44] which, as it turns out, is strongly associated with lead exposure.[45] We can play an active role in avoiding and responding to toxicity in all these areas of life.

But, unfortunately, we can't do this all by ourselves. Air and water are precious, shared resources, and it can only be through collective action that these parts of our world can be protected. If there's one thing that I think has the potential to connect us all in a world in which it often seems as though we cannot agree on much of anything, it is that the water that comes into our homes through our taps should be clean and the air that comes in when we open our windows should not be noxious, and that no one should be allowed to poison their neighbors to enrich themselves. We can all fight for this simple, sensical shared principle.

I recognize that conversations about toxins can feel very defeating. If we're marinating in hundreds of "forever chemicals" before we're even born, and swimming in thousands more as we go about our lives, and if these poisons of various sorts can wreak havoc upon our brains, then is everything else we do to protect our brainspans pointless?

No. Absolutely not. Not at all. Not one bit.

We are remarkably resilient creatures. Our brains are remarkably resilient structures. And given all that we can do to prevent brain aging in other ways—and even to turn back cognitive decline when it happens—we should not be scared into thinking that this toxic world is eventually going to come for us. It should, however, accentuate how hard we all need to work at controlling that which we can actually control.

12

THE MICROBIAL MIND

Good evening, ladies and germs.

—MILTON BERLE

Our microbiomes are so vast that it can be difficult to comprehend the extent of the lifeforms living on and within us.

Our bodies are home to tens of trillions of microbial entities, including bacteria, viruses, fungi, and archaea (single-celled non-bacterial organisms). Each of us carries thousands of different species on our skin, in our mouths, in our guts, in our lungs, and even in our brains, and these microbes can play both symbiotic and destructive roles wherever they are found.

We've known about the human microbiome for hundreds of years. Yet we've only scratched the surface of how it impacts digestion, metabolism, immune function, and disease progression. And while it has become abundantly clear that the microbes can affect our health in many ways, there remains a lot of obstinate skepticism about the role of microbes in brain aging and disease.

This reluctance to embrace connections between the microbiome and human cognition is frustrating. The old, binary way of thinking about a microbial infection—it's either acutely bad or it's nothing at all to worry about—is no longer sufficient. Today we must be cognizant of complex mixtures of various microbes that can indeed be immediately harmful but can also be benign, at first, but triggered down the road to cause vast problems to our bodies and brains.

Take, for instance, the case of Parkinson's disease. For decades the conventional wisdom was that Parkinson's is either genetic or idiopathic, meaning its cause is uncertain. That's one heck of a strange bit of binary logic—essentially "we either know how you got it, or we don't." But not everyone who is at genetic risk develops Parkinson's. So, in actuality, it's *always* been uncertain; and while there have been some usual suspects in the host of insults that might make it more likely, such as the impact of pesticide exposure on the healthy functioning of mitochondria, most cases truly have been a mystery.

Recent research, however, has suggested an important role for the gut microbiome—something that could potentially be altered to ensure that we all avoid Parkinson's. The data indicate that a single strain of bacteria may be the main culprit in the majority of cases of Parkinson's disease!

At first, this might seem to fit quite nicely into a twentieth-century model of medicine. Single-source causes are common in infectious diseases, and a majority of the successes of the past century came because health researchers identified a singular cause for a disease and, having done so, figured out a way to mitigate or eliminate it, as was the case for smallpox and polio. But we've spent a hundred years trying to apply this same approach to chronic diseases, and that has been a disaster. And when it comes to neurodegenerative diseases, it has been an unmitigated disaster. This is why I was at first suspicious of the suggestion that Parkinson's disease might have a single microbial cause.

But skepticism and willful refusal to consider evidence are two very different things. And while the research upon which that article was based isn't so cut and dried as "X causes Y," it did indeed point to a powerful possible offender: the bacterium of the genus *Desulfovibrio*, curved rod-shaped microbes that are one of the main ways that sulfur is converted into hydrogen sulfide. In the study report, a team of University of Helsinki neurologists and microbiologists described the process by which they discovered that *Desulfovibrio* bacteria enhance misshapen proteins that accumulate in clumps known as Lewy bodies, which are a marker for Parkinson's disease and the aptly named Lewy body dementia, a disorder with symptoms similar in many respects to Alzheimer's, albeit with a different pathology. There was, however, an important caveat: The research had been conducted on worms.

Usually, model organism research is a step toward further research in animals a little more like humans, such as mice, rats, pigs, or monkeys, and successes in those organisms give us the impetus to look into human trials. But what was interesting about this report was that the scientists *already* knew that humans with Parkinson's disease were far more likely to harbor *Desulfovibrio* bacteria in their guts than individuals who were cognitively normal—they'd learned this with a relatively simple census of stool samples that had been published in 2021.[1] Correlation is not causation, though, and of course the researchers couldn't just give a person without *Desulfovibrio* a big, heaping dose of that bacterium (which has also been implicated in ulcerative colitis and irritable bowel syndrome, among other diseases[2]) and sit around for years or decades to see what happens next. What they could do, quite easily, was give the human Parkinson's bacterium to worms. When they did, many of the worms quickly died—and when the scientists looked inside the brains of the survivors, they found loads and loads of the worm-brain equivalent of Lewy bodies.[3] Together, the two studies suggested that Parkinson's might be caused in large part by this one species of bacterium.

Does that mean we could eliminate it using the same strategies we've used to successfully combat so many infectious diseases? Could we essentially prevent Parkinson's by screening for the carriers of the harmful *Desulfovibrio* strains and then targeting those strains for removal? The answer is: not exactly. As indicated earlier, Parkinson's is a neurological degenerative disease, not an infectious disease. However, removing *Desulfovibrio* may end up being one part of an optimal protocol for treatment and prevention of Parkinson's, and it might help point the path toward solutions for other diseases as well. After all, the parallels between Parkinson's and Alzheimer's are striking. In both cases, antimicrobial proteins (amyloid beta and p-tau in Alzheimer's and alpha-synuclein in Parkinson's) are at play. In both cases, there is an energetic reduction (reduced oxygenation in Alzheimer's and mitochondrial dysregulation in Parkinson's). So, while the precise role of *Desulfovibrio* remains to be defined, it may turn out to be a major contributor to Parkinson's and thus to our understanding of other neurodegenerative diseases as well. In any case, it is becoming clearer that detrimental microbes are often contributing insults that accentuate brain aging and degeneration.

Though we do need additional research to better understand these connections, the microbial contribution to accelerated brain aging and neurological disease makes perfect sense in the context of what we already know about the human microbiome, particularly given the ways in which our bodies respond to single-celled organisms that our immune systems identify as potentially harmful. That response almost always includes local inflammation that, if triggered by enough malicious microbes over enough time, can spread by way of cytokines, the small proteins that travel from system to system throughout our bodies and can trigger inflammation in systems beyond that which has initially been infected, resulting in neuroinflammation[4]—which, as noted several times in this book already, is a key insult contributing to neurodegeneration. This means that an insult in the gut, for instance, could potentially affect the brain.

We also know that an imbalance of gut microbiota—too much of one little guy and too little of another—can impair the metabolization and absorption of nutrients and damage our gut lining, robbing us of the energetics our brains need to thrive and inducing systemic inflammation.

So, it's clear that these tiny organisms can have an outsized effect on our well-being, including our brain health. It's also obvious that we still have much to learn. But that doesn't mean we should just keep ignoring this vitally important aspect of our cognitive wellness. We absolutely have enough information today to help us make wise choices about protecting the healthy function of this microscopic menagerie, thus mitigating one of the key causes of accentuated brain aging and disease.

GUT CHECK

As you know by now, consuming a healthy diet is essential to maintaining a healthy brain. But even if you've faithfully committed to the tried-and-true, plant-based, mildly ketogenic diet that I recommend, if your gut microbiome is dysregulated, even the healthiest diet is of limited value. That's because the bacteria, archaea, fungi, and viruses that live in our digestive tracts serve functions just as important to our lives as our hearts, kidneys, and lungs. The microbiome is the place where many of the vitamins we need are metabolized into their usable chemical forms, essential amino acids are synthesized, carbohydrates begin the process of transforming into energy, and toxins that make their way into our bodies by way of food and water are often identified and destroyed. It has been estimated that 70 to 80 percent of all immune cells are located in our gut, and that the work these cells do in concert with intestinal microbiota doesn't just impact the ongoing battle to keep our digestive tracts healthy but, in fact, affects systemic (whole-body) immunity.[5]

As the renowned microbiologist Anne Maczulak has pointed out many times, if all the microbes in our bodies were suddenly to disappear, we wouldn't last very long.[6] And yet, when the *wrong kind* of microbes find their way into our bodies, or the *wrong mix* gets situated deep inside of us, things can go very badly very quickly—or, in the case of our brains, so slowly over many decades that we don't notice until a tremendous amount of damage has been done, a result of an inflammatory "cytokine drizzle."

Unfortunately, the microbiome is quite sensitive to even run-of-the-mill forms of inflammation. Something as simple as a flesh wound can change the composition of our gut microbiome, rendering it less healthfully balanced.[7] This is all the more reason why we must control what we can indeed control! As such, it's important not just to eat for an ageless brain, but also to eat for an ageless gut. This is to say that our diets should be high in probiotics (typically fermented foods that contain microbes), generous in prebiotics (the typically fibrous foods that also act as food for all those microbes), and conducive to the creation of beneficial postbiotics (postdigestion substances such as vitamins, amino acids, and fatty acids).

Probiotics are from fermented foods like kimchi, sauerkraut, sour pickles, miso soup, kombucha, and yogurt—foods that help maintain a diversity of species in our guts, resulting in greater cognitive function.[8] In one review of research on probiotics and cognition, twenty-one of twenty-five experiments on adults resulted in the identification of positive associations over periods from three weeks to six months (the others may simply have been too short to register an effect).[9] These connections appear to be particularly powerful when it comes to emotion, likely because bacteria in the gut can produce dopamine, serotonin, and norepinephrine, all of which are known to have profound effects on human mood—which, as I've previously noted, is often treated as a separate domain from cognition, even though it really shouldn't be.

It doesn't matter much if we keep wolfing down "good guy" bacteria if there isn't enough food down there for them to do what they're supposed to do. Like any other organism, without proper food, microbes will become dysfunctional and die. It turns out, though, that the best prebiotic foods, by far, are those that are high in fiber: roots, greens, grains, and seaweed. These "microbe fuels" also just so happen to be foods that fill us up for longer, permitting for the fasting periods and caloric restriction we know to be helpful for lifelong brain health.

When our intakes of probiotics and prebiotics are working well, the postbiotic effects generally take care of themselves, and the result should be healthy levels of microbe-produced vitamins, especially the B vitamins and vitamin K_1 and K_2. Thiamine (B_1) is known to play a crucial role in energy metabolism, neuronal function, and cognitive development.[10] Riboflavin (B_2) has been shown to mitigate cognitive impairment via its antioxidant and anti-inflammatory properties.[11] Niacin (B_3) has been shown to have a very distinct linear relationship with cognitive function in older humans.[12] Deficiencies of pantothenic acid (B_5) have been shown to be associated with neurodegeneration and dementia.[13] Adequate pyridoxine (B_6) is known to be associated with substantially better performance on cognitive tests.[14] Copious research on folate (B_9) indicates it plays key roles in decreasing homocysteine, improving vascular health, attenuating inflammation, and enhancing antioxidation, all of which lead to improved cognition.[15] By simply raising levels of cobalamin (B_{12}) in people suffering from the early stages of symptomatic cognitive impairment, one group of researchers reported they were able to improve cognitive test results in three-quarters of the individuals they tested[16]—an important finding given that cobalamin (B_{12}) insufficiency is so common, affecting about 13 percent of US residents over the age of nineteen.[17] Multiple studies have demonstrated direct and substantial correlations between low concentrations of phylloquinone, also known as vitamin K_1, and deteriorated cognitive health.[18] These are all vitamins that can, and in many cases should, be obtained through food and, when necessary, supplements. But since many people have trouble absorbing these vitamins when ingested, the fact that our gut microbes can also produce them is a good reason to keep working toward probiotic and prebiotic intakes that produce the Bs and Ks as a postbiotic outcome.

MOUTHY

One of the major trends of the past several years is the recognition that oral health is closely tied to cognitive health. Gingivitis, periodontitis, root canals, occult abscesses, dental amalgams, and herpetic cold sores all play important roles in risk for cognitive decline. Thankfully, there is an increasing number of sophisticated, highly trained dentists and oral-systemic specialists who are expert at diagnosing and treating these conditions. In our clinical trials, we have worked with such specialists to ensure detection and rectification of these conditions. Oral DNA testing often reveals pathogens such as *Porphyromonas gingivalis* and *Treponema denticola*, which are associated with cognitive decline, and which can be eradicated. Oral probiotics and prebiotics are increasingly available and effective, as well as rinses for reducing pathogens, such as StellaLife and Dentalcidin, and the Ayurvedic practice of oil pulling. Cone beam analysis may reveal occult abscesses, and, in the absence of treatment, these can induce long-term inflammation and compromise cognition.

When this oral-neural connection was first identified, it was assumed that dental problems—from tooth decay to cavities to gingivitis to halitosis—were a tertiary symptom of cognitive decline, since people who were having trouble with their memory were more likely to forget about oral hygiene. As the years have gone by, though, it's become increasingly clear that dental problems typically precede symptomatic cognitive decline, and the mechanisms mediating these interactions have emerged.

Pathogens, toxins (such as mercury), and reduced energetics (in those with obstructive sleep apnea) are to blame. While the oral cavity is home to many commensal bacterial species, it's also where you'll find some harmful species, most notably *Porphyromonas gingivalis,* another rod-shaped bacterium that, if allowed to live in a person's mouth for too long, can cause serious infections and lead to inflammation. And while we know that inflammation anywhere in the body can ultimately lead to neuroinflammation, it's worth remembering that the oral cavity is *really* close to the brain.

It's not just inflammation that we have to worry about. *Porphyromonas gingivalis* also expels neurotoxic enzymes called gingipains that researchers have identified in more than 90 percent of the brains of people who have died of Alzheimer's.[19] And these enzymes have also been shown to increase the permeability of the blood-brain barrier, providing even more avenues for other toxins to make their way to our brains.[20] *Porphyromonas gingivalis* is truly a curse that keeps on cursing.

Fortunately, oral health is something almost everyone can improve upon through daily brushing, flossing, and rinsing; regular dental care, dietary choices (chiefly avoiding sugar), and oral probiotics, especially those that can increasingly be found in toothpastes (such as Revitin and Dentalcidin) that have been specifically designed to support a healthy community of oral microbes.

In just the past few years, I've been impressed by how quickly dentists have wrapped their arms around the key role they play in fighting against accentuated brain aging and neurodegeneration. At a time in which many physicians remain reluctant to speak with their patients about proactive measures they can take to prevent cognitive decline, dentists like Howard Hindin are often filling a crucial niche. Hindin, cofounder of the American Academy of Physiological Medicine & Dentistry, has been in practice for more than fifty years and has never stopped looking for ways to help his patients stay healthy. So, as he learned that oral health was so influential to long-term brain health, he was eager to share that information with the families he serves.

And these don't have to be serious and scary conversations. Hindin has a folksy charm that he employs to get his patients to think about more than just keeping their teeth clean and preventing cavities. "My uncle had a Chevy that twenty-five years later looked, sounded, and ran like it just came out of the showroom. It took time, energy, and consistent diligence," he recently told me. "With the same effort we can keep our brains healthy for a lifetime by seeking and treating the causes rather than the expressions of problems."

I couldn't have said it any better! If we're to stay brain healthy for a hundred years or more, we need to tend to every part of our body, and especially our mouths.

AN INFECTIOUS IDEA

About a quarter of the global population carries at least one copy of the APOE4 allele, and about 2.5 percent of the population carries two copies—a genetic inheritance that has recently, albeit incorrectly, been communicated as an inevitable ticket to suffering and death from Alzheimer's.

It's interesting to consider how gene variants like this become so prevalent across a population. After all, if the variant is really so detrimental, it's not creating individuals who are fitter for survival, and thus it might seem reasonable to expect variants like this to dissipate over time, rather than to be so common that virtually everyone in the world knows many people with multiple copies of this gene sequence. This makes more sense when you stop thinking about APOE4 as an "Alzheimer's gene" and begin thinking about it as a "hyperimmunity" gene. For many of our ancestors, a more aggressive immune response and the inflammation that comes with it would have been a *benefit*, as it would have prioritized healing from infections early in life—before and during fertility—over the long-term consequences that come from too much inflammation, which generally come along after parenting has commenced and young offspring have had a chance to grow and begin families of their own.

But it's not just people with two copies of APOE4 who enter their forties, fifties, and sixties having battled a lifetime of inflammation and thus become more susceptible to neuroinflammation. It's also those who were simply exposed to more pathogens. Thus, even if you're not more genetically susceptible to neurodegeneration via a gene that heightens your immune response, you might still face a heightened risk if your body has had to launch its defenses, again and again, in response to infections. Put simply, more infections usually mean more inflammation, and more inflammation usually means more neurodegeneration.

This is why it was not surprising to learn that when researchers reviewed the records of more than 6 million people during the first year of the COVID-19 pandemic in the United States, they saw clearly that those who had gotten COVID were at significantly increased risk for a new diagnosis of Alzheimer's within the next year.[21] Alas, it would appear that people whose brains had been running on the edge for a very long time simply couldn't take one more infectious insult.

I've seen examples of this repeatedly—people whose reservoirs of cognitive resilience appeared to have been emptied by an acute infection. One who immediately comes to mind, because her case was so stark, is Abigail. Like many people in our world, Abigail had survived multiple bouts with COVID, and she had begun to think of the disease as little different than a common cold. But the fifty-three-year-old construction supervisor had a very different experience with her third case. "It knocked me down hard," she said, "and I had a lot of trouble breathing, just walking up the stairs to my bedroom, all those things that I'd definitely seen other people go through, but I guess I just thought wouldn't happen to me because I had always been in pretty good shape."

Those symptoms subsided, but the brain fog did not, and Abigail had lapses in memory and difficulty focusing. She thought it would eventually go away, but a year later she was still struggling and had to take a leave of absence from her job. A MoCA test indicated mild cognitive impairment while a blood test showed an increased level of phosphorylated tau, akin to what we might see in early dementia. A slightly increased neurofilament light chain test suggested that neurodegeneration had indeed commenced.

As of this writing, it has been about a year since her symptoms began, and I'm thankful that a range of interventions has helped reverse many of them. Abigail says she no longer feels foggy. Her MoCA score has normalized, and she has returned to work. Troublingly, her p-tau 217 and NfL tests have shown only minimal improvement, but at least they have not worsened. All things considered, she is on the right track.

But what is frightening about Abigail's case is just how common it is. In 2024, for instance, a team of Brazilian researchers investigated the cases of sixty-three people, aged eighteen to fifty-nine, who had experienced brain fog after a case of COVID, a result of the cytokine storm that happens as a result of this particular virus and others that cause acute disease.[22] Even when the initial sickness was reportedly mild, the researchers found that the degree of cognitive loss and exacerbation of fatigue were strikingly correlated with higher NfL levels.[23] These patients had brain damage that looked in many respects like that which often takes decades of very slow neurodegeneration to achieve.

I share this not because I want to make you paranoid about COVID. It is, alas, part of our world and will be for a long time to come, and it's important to realize that most people who have gotten it *don't* have symptoms like these. Why Abigail and others suffer in this way is a question we'll be working to answer for years to come.

What I do want to drive home is how powerful an insult a pathogenic infection can be on our cognitive health, either acutely or cumulatively over time. Infections aren't something to pass off as no big deal. To the extent we can avoid them, we obviously should. Antibiotics should not be overprescribed because they can disrupt and damage the gut microbiome by killing beneficial bacteria.

Well-tested vaccines can be a substantial part of this strategy. Influenza, pneumonia, tuberculosis, and shingles (HVZ) vaccines have all been shown to reduce the risk of neurodegenerative disease, likely because by preventing or diminishing these infections, they lessen the lifetime burden of inflammation, and thus, neuroinflammation.[24] The same is true of vaccines for bacterial infectious diseases like tetanus, diphtheria, and pertussis.[25] Even when we control for all the most substantial cofounders for disease, we see compelling evidence that by reducing acute infection and the cumulative lifetime burden of infections through vaccination, our brainspans benefit.

At the time of this writing, we simply don't yet have enough evidence to know whether the COVID-19 vaccination also offers this sort of long-term cognitive protection. We do know that it offers broad protection against the virus's most adverse *immediate* outcomes, such as cytokine

The Microbial Mind

storm, pulmonary failure, and death. However, we have seen some people begin or accelerate their cognitive decline following both COVID infections *and* COVID vaccinations—perhaps because vaccines trigger some degree of inflammation due to their adjuvants (immune response enhancers). The association between COVID and increased incidence of cognitive impairment and Alzheimer's diagnoses in the three months postvaccination was also described by a team of researchers from the United States and South Korea who called for continuous monitoring and investigation into the vaccines' long-term neurological impacts.[26] As always, the goal is the same: We want the best outcomes. So, we will be watching this closely. In the meantime, reports like this highlight a key principle about cognitive health: The closer we permit ourselves to approach the cognitive edge, the easier it is for something that, in most cases, would be a very minimal insult—even something from which we might normally derive a great benefit—to push us over.

SEXUALLY TRANSMITTED INFECTIONS AND BRAIN AGING

There are many potential sources of inflammation that may increase brain aging, such as periodontitis, leaky gut, mycotoxins, viruses like COVID-19, and ones that are on the rise globally: sexually transmitted infections (STIs).

In recent years, protected sex has fallen among all demographic groups. According to one report, only about a third of women and a little more than a quarter of men who engage in sexual activities that put them at risk for infection use condoms.[27] As a result, sexually transmitted infection rates have been surging in ways that Hilary Reno, a professor at the Washington University School of Medicine in St. Louis, called "stunning." "I can't tell you how many primary-care physicians have called me recently and said, 'I just saw my first-ever case of syphilis this year,'" she lamented in 2022.[28]

Because treatment and prevention had rendered syphilis so rare for a long time, most people don't know that the infection was, for most of its epidemiological history, associated with a type of madness called general paresis, which is caused by the atrophy of the brain's frontal and temporal lobes. Today, although not unheard of, syphilis is generally regarded as an uncommon cause of dementia.[29] I suspect, however, that we will come to see that the recent resurgence will be followed by some number of years with an increased rate of symptoms of cognitive impairment among those who acquired the disease, perhaps even in cases in which it was quickly treated but especially if it was not immediately addressed upon the first signs of infection.

The most common sexually transmitted infections include a number of curable diseases like chlamydia, gonorrhea, syphilis, and trichomonas, and several incurable but treatable conditions like herpes simplex virus, HIV, and human papillomavirus. It should go without saying that even if treatable, and even if curable, these are not infections anyone would want to have. But more important, these and other infections—including herpes, chlamydia, and spirochetes—are likely significant contributors to brain aging and dementia. "Some microbes can remain latent in the body with the potential for reactivation, the effects of which might occur years after initial infection," more than thirty Alzheimer's researchers argued in an editorial in the *Journal of Alzheimer's Disease* in 2016. "These agents can undergo reactivation in the brain during aging, as the immune system declines, and during different types of stress." The consequent neuronal damage happens again and again, the researchers wrote, either directly leading to or at least acting as a cofactor for synaptic dysfunction, neuronal loss, and, ultimately, dementia.[30]

Nearly a decade after that editorial, there is still a woeful lack of research on this aspect of accelerated brain aging and disease, which is one of the reasons why most people haven't heard of this likely connection. But I believe they should. Just as dentists have taken the onus of talking to their patients about long-term cognitive health, a responsible sexual education curriculum should include frank discussions about the potential consequences of sexually transmitted infections on brainspans.

Sex is an important part of life and it is certainly not my place to tell people when and how to have it. What I will stress, however, is that *safer* sexual practices aren't just about protecting ourselves but protecting our partners—potentially for the rest of their lives.

TICKED OFF

Back in 2001, the epicenter for Lyme disease in the United States was Lyme, Connecticut, as part of a roughly thirty-thousand-square-mile region stretching from northern Virginia to southern Massachusetts along the eastern seaboard, where deer ticks have been carrying the bacteria that causes the disease *Borrelia burgdorferi* for sixty thousand years.[31] During all that time, the ticks had more or less stayed put.

But in the past two decades, the Lyme disease epicenter has grown to some 160,000 square miles, spreading north all the way into Maine and west across all of Pennsylvania and New York state. What's more, an area that had previously had only a small number of cases, in and around the greater Minneapolis area, has become its own epicenter for Lyme, having spread to encompass nearly all of Minnesota and Wisconsin. And as of today, temperate climes around the world, including Japan, Canada, and countries in Europe, are home to various species of *Borrelia*, the Lyme spirochete. And the parallels between chronic neurosyphilis and chronic neuroborreliosis—each produced by a spirochete—are of grave concern. Accordingly, in our patients with cognitive decline or risk for decline, we always check for evidence of tick-borne co-infections, including *Borrelia*, *Babesia*, *Bartonella*, *Ehrlichia*, and *Anaplasma*. Unfortunately, the current diagnostic tests are relatively insensitive at determining whether these chronic infectious agents are present, in part because there are many different species, only some of which are included in the available tests.

As a global community, we appear to be close to, if not just past, the peak of the emissions of greenhouse gasses that trap heat inside the lower levels of our atmosphere.[32] Even then, however, global temperatures are likely to continue to stay well above pre-industrial averages for decades to come as CO_2 remains in our atmosphere.[33] It turns out that the Lyme-carrying deer ticks like forests and grasslands near forests, where their deer hosts live, and some scientists believe that climate change may increase their territorial reign of terror.

Lyme disease is typically treatable with a short course of antibiotics, but much like COVID, it is also known to sometimes and seemingly randomly cause prolonged fatigue and cognitive dysfunction—symptoms that may be explained by a study by researchers at the Johns Hopkins University School of Medicine, who used novel brain-scan technology to show that twelve people who had such symptoms after contracting Lyme disease all had elevations of a chemical marker for widespread brain inflammation, while the nineteen controls had hardly any at all.[34]

For a while, a few years back, there was quite a bit of speculation that the same bacteria that causes Lyme disease may cause Alzheimer's disease, and some people went so far as to suggest that the diseases might even be one and the same. That's not the case, because, of course, there is no *single* cause for neurodegeneration. However, that doesn't mean tick-borne infections aren't playing a role in accelerating neurodegeneration—clearly, they can do that. Indeed, others will argue "that's just Lyme, not Alzheimer's," but this ignores the numerous other tick-borne pathogenic diseases that may contribute to cognitive decline, including babesiosis, anaplasmosis, ehrlichiosis, Powassan virus, tick-borne typhus, tick-borne encephalitis, and Rocky Mountain spotted fever, that have been shown to breach the blood-brain barrier and accentuate inflammation. Even if it was "just Lyme," though, I'd be no less concerned—we always need to identify and treat those contributors to our brain aging and disease so that we can optimize outcomes.

Julie is a good example of this. She was forty-four years old when she was bitten by a tick and contracted Lyme disease. Five years after treatment for that disease, her symptoms of cognitive decline began. Fortunately, unlike many people, she didn't wait for those symptoms to worsen before seeking help. The lifestyle interventions we shared with her did the trick, and for five years she felt as though her brain was "good as new."

But at fifty-four, she started going backward, and we learned from further testing that while she didn't appear to harbor lingering markers of Lyme disease, she did have babesiosis, likely another result of the same tick bite, since more than 50 percent of people who develop Lyme also develop one of Lyme's co-infections. Moving forward, she will continue to be vigilant, which is an important part of any winning strategy for cognitive health, but especially for those who have had tick-borne illnesses.

It's important to take precautions against ticks. This means wearing close-toed shoes, long-sleeved shirts tucked into pants, and long pants tucked into socks or boots when in tick-infested areas. Insect repellent is a good idea as well. Finally, check your body, including your scalp and hair, after spending time in tick territory and, if one has hitched a ride on your skin, remove it immediately. If you have been bitten, you don't need to panic; use tweezers to pull the tick out from as close to its head as possible, wash the location with soap, and then follow that with rubbing alcohol to disinfect the bite site. Lyme is generally only transmissible if a tick has been embedded for more than twenty-four hours, but there is less certainty about how quickly other pathogens can be transmitted, so if you know that you've been bitten, or if you live in tick territory, it may be worthwhile to send the tick to a lab for testing or occasionally test for markers of these pathogens. Most important, remember that ticks carry many other pathogens in addition to *Borrelia*, so please get tested for these others as well, so that any needed treatment will not be delayed. As noted above, more than half of people who contract Lyme also become infected with a Lyme co-infection, such as *Babesia*.

THE INFECTIOUS WORLD

As of this writing, there isn't a broadly available way to test for the most common pathogens associated with tick bites. Worse, we're now learning that the pathogenic infections that may be implicated in neuroinflammation are likely not just tick-borne. We've known for years of the "hidden burden" of malaria—cognitive impairment follows infection by protozoa of the genus *Plasmodium*, which is transmitted by the bite of the female anopheles mosquito.[35] And while we may finally be on track to eradicate malaria, perhaps by the middle of this century,[36] there are other mosquito-borne diseases that are present and growing around the world, including West Nile virus, Zika virus, La Crosse encephalitis, chikungunya virus, dengue, and yellow fever. Then there is cat scratch fever, lymphoreticulosis, which is mainly caused by bacteria called *Bartonella henselae*, but there are many *Bartonella* species that might cause this disease as well.

I could go on and on. By some estimates, there are at least a trillion microbial species out there, and if each individual microbe was laid end to end, they would stretch out for more than 100 million light-years.[37] If you have trouble fathoming these sorts of distances, just consider this: The nearest star to our sun is only about 4.25 light-years away! Obviously, not all those organisms will find their way into our bodies, but a lot of them will. There are about 40 trillion individual microbes in your body right now. But of all the species out there—and *in* there—only a small minority, about 1,400 or so, are known to be a cause for infectious disease.[38] There are likely many more, but it's nearly impossible, at this point, to even test for the microbes we already *know* to be disease-causing.

So, at least for now, I consider microbes in the same space in which I consider the many neurologically harmful toxins in our world—as an inevitable insult that will come for most, if not all, of us at some point. But I also take a lot of comfort in how fast we are moving toward a world in which we will be able to rapidly detect thousands upon thousands of disease-causing microbes in the body. While he is better known for his work in the field of the epigenetics of aging, Harvard Medical School geneticist David Sinclair has also been at the forefront of this effort, and the high-throughput genetic sequencing technology—which began in his garage after his daughter contracted Lyme disease and had to wait a week for lab results as she suffered from severe and worsening symptoms.[39] His is just one of many efforts across the globe to make it easier to identify potential pathogens and thus pathogen-specific solutions, since there is simply no one-size-fits-all solution for every harmful microbe.

Soon enough, I do believe, we'll have the technology to quickly and cheaply identify and target virtually all the specific microbes that lead to the many chronic infections associated with cognitive decline. In the meantime, though, there is plenty we can do. Typical testing for these various pathogens utilizes antibody testing and FISH (fluorescent in situ hybridization, which detects the nucleic acids from the pathogens).

We can nourish our gut microbiomes with healthy diets, prebiotics, and probiotics. We can double down on oral health. We can take reasonable precautions to avoid transmissible viruses and make informed choices about vaccines that have been demonstrated to not only prevent acute illnesses but, in doing so, protect long-term cognitive health. We can engage in safer sex. We can protect ourselves from ticks. With each of these steps, we buy ourselves protective space and time, such that we are more likely to benefit from the technologies and interventions that are on the way.

13

SOONISM

> People who say it cannot be done should not interrupt those doing it.
>
> —COMMONLY ATTRIBUTED TO CONFUCIUS AND GEORGE BERNARD SHAW, BUT NEITHER ACTUALLY SAID IT

By now, you probably have recognized that there are many aspects of your life and lifestyle that may not be perfectly aligned to having the best cognitive health today and ever more. I understand that might feel discouraging, so I want you to know something very important: You don't have to solve everything *right now* to have a healthy brain at one hundred years old.

Of course, a journey toward neurological wellness that begins today is better than one that begins tomorrow. But it's not easy to make and maintain lifestyle choices consistently that support brain health, which is why people who try to do anything and everything, all at once, are often unsuccessful.

Here's the very good news: If you're cognitively healthy now—even if you have thus far not taken all the actions we've discussed throughout this book to prevent brain aging and disease—you almost certainly still have time to make the necessary changes one by one. That's true whether you're twenty years old, forty years old, sixty years old, or eighty years old.

I know that might sound wildly optimistic, so let me explain.

Let's say you're eighty and, as it happens, you've made it this far with your brain still doing everything it really needs to do. You can remember the important things in life. You don't get easily lost or confused. You can focus on the tasks that need to be done. You're a generally happy person. Chances are good that you've lived a decently brain-healthy life even if that wasn't your explicit intention. You've maintained a reasonably good diet. You've exercised throughout your life. You've encouraged brain plasticity by learning new things and staying social along the way. Somehow, you've managed to avoid too many toxins and infections. It's also likely that, if you've gotten to eighty without substantial brain aging or disease, you've gotten a little bit of help from the genetic lottery. For reasons that you cannot take credit for, but can certainly feel blessed about, you're not predisposed to neurodegeneration. None of this means there hasn't been *any* age-related cognitive decline, but if it is not yet diagnosable as anything *more* than age-related cognitive decline, you've done all right for yourself. From here, a brainspan of one hundred isn't hard to envision, for you've made it most of the way to that goal already, and there are only twenty years to go.

Now, it must be said that just staying the course is almost certainly not going to be enough. Another twenty years of very good brain health is unlikely without some intentional and relatively substantial changes. Your diet will likely need to get better. Your exercise regimen may need to improve. You will need to engage in some intentional brain training. You will need to manage your sleep. These aren't easy steps to take at eighty years old, but these are also not *impossible* steps to take, especially if taken one by one with the goal of building capacity that can be used to achieve the next step.

If, by your eighty-first birthday, you've transformed your diet into one that is plant-based and mildly ketogenic with at least twelve hours of fasting and adequate nutrients to provide the energetics your brain needs, then in that single year you've likely made cognitive gains that will help you get to eighty-two. If, by your eighty-second birthday, you're exercising most days with a combination of aerobics, strength training, and perhaps a bit of blood-flow restriction and exercise with oxygen therapy, those actions have likely bought you enough of a brain boost to get to eighty-three. And if, by your eighty-third birthday, you've made a habit of daily, monthly, and annual cognitive challenges, while doubling down on developing the social interactions that support neuroplasticity, you're probably going to make it to eighty-four with an even better brain than you've had for a very long time. Keep making changes, adjusting and optimizing with intentionality, guided by the best available science, and you'll be well on the path to becoming a cognitive guide star for others, with an exceptionally healthy brain at one hundred years old or whatever your natural lifespan turns out to be.

Can you move more quickly toward these goals? Almost certainly. Should you? If you're able to, of course! But the point that I'm making here is that, if you've made it this far in decent cognitive shape—even if you're eighty years old—you don't have to go into panic mode in an attempt to reach a hundred-year brainspan. Steady progress is the key.

Now, it might seem obvious that if you can make these sorts of changes at eighty and get to one hundred with a healthy brain, then there's nothing really to worry about at the age of twenty. You can worry about this later, right?

But this is where things get a little tricky. Because if you've made it to eighty with a very functional brain, you've gotten lucky in some respects, and you only have to build upon that luck for twenty more years. If you're twenty, though, you've got *a lot* of life ahead of you—including a lot more time in a world of toxic exposures potential infections. It's eighty more years of work and family stressors you can't see coming.

And, if you happen to have a genetic predisposition, it's a lot of time for the odds to accumulate against you.

Thus, another eighty years of very good brain health is, unfortunately, quite unlikely without some substantial efforts on your part to support that goal. But if, by twenty-one, you're eating a brainspan-protective diet; and if, by twenty-two you're working out the right way every day, and if, by twenty-three, you're getting the intentional cognitive exercise your brain needs, too, then you're also moving in the right direction to reach one hundred years with a young and healthy brain, for these are investments that will pay dividends for the rest of your life.

I note this with confidence because there are also a lot of new protective strategies and therapies that will come along in the years ahead of us. How quickly you embrace these—the threshold of evidence that you will require to add something to your brain wellness regimen—is your decision. It will depend on your own circumstances, the advice you're getting from a trusted doctor, and the way you view potential risks and rewards. I won't tell you how to make that decision, but I will say that most people are at least a little bit risk averse—and that's not necessarily a bad thing.

For example, many of us have been taking supplements to increase the coenzyme NAD+ (such as nicotinamide riboside or nicotinamide mononucleotide), a very promising molecule to support energetics. Indeed, NAD+ and its precursors may be the most popular supplement in the world for those seeking to reduce their rate of aging in general and brain aging in particular. Because of this, it's also among the most well researched, and yet even for this molecule, the research—still mostly in animal models—remains nascent and the results mixed, with plenty of reports of null or adverse effects.[1] Yet I know a lot of physicians who, having read the literature, considered the risks, and thought about the potential benefits, have advised their patients that a NAD+ precursor is a good bet for staving cognitive aging.

Does that mean *you* need to be on NAD+ *right now*? Perhaps not.

But what it does mean is that there are lots of promising interventions that are not yet mainstream but will likely be much better understood in the future. In other words: You're going to get a lot of help along the way!

Now, I do love science fiction. And I can happily ruminate on what might be in store for us fifty years down the road. But, of course, the further out we go, the more speculative those ruminations get. So, instead of being futurists in this chapter, let's be *soonists*—let's talk about the strategies and therapies that are either available now or are very likely to be ready *soonish* and are likely to better the odds that anyone, at any age, can remain free of brain aging and disease for one hundred years or our lifespans.

BIO-IDENTICAL HORMONES

Hormones—the body's ubiquitous messenger molecules—are essential to every part of life. They affect blood sugar and blood pressure. They impact metabolism and sleep. They drive our desires and moods.

It should come as no surprise, then, that physicians like Ann Hathaway, a family practice physician who specializes in functional medicine, Prudence Hall, an OB-GYN specializing in bio-identical hormone replacement therapy (BHRT), and Felice Gersh, an OB-GYN and integrative medicine specialist, along with many other physicians, are so bullish about the use of bio-identical hormones (those that are chemically identical to the ones created in our bodies, as opposed to synthetic hormones that are structurally different from our natural chemical messengers). Bio-identical hormones like estradiol and progesterone can be a powerful component of the best precision-medicine approach to treating cognitive decline.

We know this in no small part because, while hormone replacement therapy (HRT) continues to be a fringe treatment for some conditions, including aging, there has been tremendous progress over the past few decades in using hormone replacement to address the symptoms of menopause. In most cases, cognition wasn't the only target for physicians who have prescribed hormones to help women who are experiencing symptoms before, during, and after menopause, but as many women will frustratingly attest, one of the most common of those symptoms is brain fog.

That's no coincidence, of course. Healthy levels of estrogen and progesterone are well known to be connected to memory, spatial awareness, problem-solving, and fine motor skills. What we've seen over the years is that these particular neurological symptoms are often greatly alleviated by BHRT.

The troubling caveat: Some studies have indicated that women who have used HRT appear to be more likely to suffer from dementia later in life. Other studies have shown the opposite. How can this be? Well, these associations are very complicated. Women experiencing a greater degree of negative symptoms are more likely to seek and use HRT, thus there has been some debate as to what is actually driving the associations.

Recently, though, a research team led by Lisa Mosconi, the director of the Alzheimer's Prevention Program and the Women's Brain Initiative at Weill Cornell Medical College in New York City, identified a possible answer for why some studies show that HRT inhibits dementia while others show it may be associated with increased rates of age-related neurological diseases. It turns out that there may be a "sweet spot" for treatment. In an analysis of tens of thousands of people enrolled in randomized, controlled trials and hundreds of thousands of patient records, Mosconi's team discovered that HRT is most effective at protecting brain health when taken as close as possible to the point at which menopausal symptoms begin and continued for at least ten years. In these cases, the risk of dementia was reduced by nearly a third.[2] Another study of nearly 400,000 women found that BHRT was more effective than nonbio-identical hormones, and that the risk reduction included multiple neurodegenerative conditions, not just Alzheimer's.[3]

This finding aligns perfectly with our understanding of brain aging and disease as problems that occur as a result of an accumulation of insults. When we address insults as close as possible to when they occur, we can prevent future problems.

It's obviously critical that women understand this, but these findings are important for everyone, because there continues to be a dearth of research about how supplementation and replacement work in other areas of wellness. Nonetheless, in recent years there has been a very noticeable increase in physicians who, informed by their patients' needs and data, prescribe bio-identical hormones both for menopause-related cognitive symptoms *and* for the prevention or as an intervention for cognitive decline, whatever the reasons for that decline might be.

Many others remain on the sidelines, waiting for "more research" that may never actually materialize, not because it can't be done but because there is little profit in it, since drug companies can't get a patent for bio-identical chemical structures. And while I do wish to emphasize that a conservative approach to medicine is not necessarily a bad thing, how long shall we wait? Given how important hormones are in our bodies, should we not desperately be seeking to understand how the changing levels of these chemicals in our bodies impact our health over a lifetime and how replacement with bio-identical hormones might help us live longer and healthier lives? If you talk to the experts like Dr. Hathaway, Dr. Hall, or Dr. Gersh, they will tell you that they often see cognitive improvement in their patients taking BHRT.

Thus, in the vast majority of cases, well-meaning physicians are left using what they know from the available research, and what they suspect will work based on their understanding of their patients, to make cautious decisions about hormone supplementation. All of this is why I've chosen to categorize at least these parts of hormone optimization as a soonist strategy for brain health—because even though we should know more by now, there is still some controversy involved in hormone treatment for aging.

There are at least sixty different hormones created in the glands spread across our bodies as far north as the hypothalamus and as far south as the ovaries and testes. While it might seem simple enough to take measurements of these, identify those that are "low" versus some standardized estimate of what is healthy, and replace them with bio-identical chemicals, in practice everyone's internal milieu is quite different, and there are many hormones that simply haven't been well researched, especially when it comes to brain health. What's more, hormone imbalances are often self-correcting when other brainspan interventions are taken. A brain-healthy diet provides the nutrient substrates for the creation and function of signaling chemicals.[4] Exercise is well known to amplify the creation and activity of hormones.[5] The same is true for improvements in sleep, decreases in negative stress, and the clearance of toxins and infections. Thus, most people need not *start* with hormone supplementation.

That absolutely shouldn't suggest that I think hormone supplementation, toward the goal of optimization, is bad. But for most people, I believe, bio-identical hormone replacement should play a role similar to the proper role of pharmaceuticals—as an intervention that can be used as part of optimization, and when immediate recovery of a hormone is imperative. The upshot of this is that, outside of menopausal women, most people don't need hormones *right now*, which means most of us can benefit from the experiences of people and physicians who have chosen to be vanguards.

We do have a reasonable amount of experience at this point to make better-informed personal decisions, in consultation with a physician we trust, on four hormones in particular.

Melatonin is most commonly associated with its role in sleep, but this free-radical scavenger has also been shown to have strong neuroprotective effects, especially against neuroinflammation,[6] and research tells us that melatonin levels in humans are inversely associated with cognitive deficits—the more deficient in this hormone we are, the more brain aging and disease we tend to have.[7]

Human growth hormone, also known as somatotropin, is most associated with the stimulation of cellular growth—hence the name—as well as cell reproduction and regeneration. These are, of course, attributes that are as important to neurons as any other cell in the body. Decreases in growth hormone levels are associated with reduced executive function and lapses in short-term memory as well as increased rates of Alzheimer's disease.[8] Some choose to increase growth hormone levels by taking oral secretagogues, which drive the pituitary to produce increased amounts of growth hormone, although the effects of secretagogues may wear off after six to twelve months.

Most people seeking testosterone supplementation use this hormone for its beneficial effects on energy, muscle strength, and sexual function, but we also know that "low T" is strongly correlated with higher prevalence of cognitive decline and dementia.[9]

And DHEA—dehydroepiandrosterone, an important precursor to testosterone and estrogen—is most often used with the intention of strengthening immunity, increasing energy, and improving sex drive, but when it drops, cognitive decline may also follow.[10]

There are many other hormones that likely play a role in cognitive health, including the main thyroid hormones like triiodothyronine and thyroxine as well as progesterone, pregnenolone, and cortisol. For the best outcomes, we must ensure that all these are in an optimal range. More research is underway, so that soon, we'll have much more data upon which to make decisions about the best ways to maintain hormonal balance, and whether those methods result in slower cognitive aging and less neurodegenerative disease.

SENOLYTICS

"Zombie cells" is a term often used for cells that have gotten old and stopped functioning normally but have not died. They have become senescent cells, and it has been speculated that this phenomenon prevents the cells from becoming cancerous. However, these cells also send out chemical signals, part of the senescence-associated secretory phenotype (SASP) discovered by biochemist Judith Campisi. Chemical signals may function to call in new replacement cells from nearby stem cell depots, but whatever the purpose, the SASP includes inflammation, and removing the senescent cells seems to help reduce aging. (Research on brain effects is ongoing.)

Therefore, therapeutics that destroy the senescent cells—senolytics—have been developed, and show promise in extending lifespan and brainspan. Interestingly, fasting is one method to induce this senolytic effect. There are also supplements such as fisetin (300 milligrams per day) and quercetin (500 milligrams twice per day) as well as the drug dasatinib, which all exert a senolytic effect. Remarkably, senolytics also increase the production of an antiaging molecule called klotho,[11] which is associated with enhanced cognition.

PEPTIDES

I'm not sure a day has gone by over the past few years in which I haven't been asked about peptides—the short strings of amino acids that include some hormones but also include other signaling and structural molecules. Because of their small size, peptides are easily absorbed and integrated into our bodies and more readily traverse the blood-brain barrier. Hundreds of different peptides, with many different functions, are now available, and the relative dearth of quality human studies hasn't stopped many people from experimenting with them. They're thought to have cognitive benefits, and an increasing number of physicians are helping guide their patients into and through that experience.

I'm broadly in favor of greater health freedom—not unfettered access to anything, to be certain, but more room for doctors and their patients to make their own decisions. That said, a lot of people seem to think peptides are a shortcut or cure-all for brain aging, and that's simply not the case. The first and best strategy for optimizing peptide creation is to let the other interventions we have discussed in this book take a natural effect.

I'm very much open to the potential for peptides to have a substantial role as supplemental interventions in a person's long-term brain health regimen guided by a physician.

Among the most popular peptides intended to improve cognitive health is dihexa, which appears to improve learning and other mental functions in rodent studies,[12] but which has not been well studied in humans, despite its growing popularity and use by some doctors who believe it may have potential to slow the progression of Alzheimer's. Another peptide that is growing in popularity, particularly with those who have a lot of disposable income since it is prohibitively expensive for most, is PE 22-28, which has shown similar promise in rodent studies and which some people have reported to have substantial antidepressant effects but is not known to be effective or safe in humans.[13] Much less expensive—and thus quite a bit more popular and growing more so—is the peptide pinealon, which appears to help rats remember learned skills, such as running mazes,[14] but has not yet been studied in humans enough that we understand its therapeutic effects or side effects. Epitalon is another synthetic pineal-derived peptide, and it has been shown to increase telomere length, which has been associated with antiaging and reduced inflammation, albeit only in specific subsets of patients.[15]

Two other popular peptides, thymosin α1 (TA1) and thymosin β4 (TB4), are derived from the thymus gland, a key site for cellular immunity. TA1 enhances the immune response to pathogens that enter cells, such as viruses and some bacteria. As we age, our specific immune responses typically decline, and the nonspecific inflammatory component increases, so improving targeted immune responses is a promising avenue to long-term brain health. TA1 is usually given by subcutaneous injection twice per week, at a dose of 0.8 to 6.4 milligrams.[16] TB4 functions to support tissue regeneration and thus holds special promise for those who are rebuilding brain synapses;[17] the dosage is typically 0.25 milliliters injected daily.

One of the peptides I've become quite interested in is davunetide, which is an octapeptide, meaning it has eight amino acids, derived from the sequence of a highly potent neurotrophic factor known as activity-dependent neurotrophic peptide (ADNP). It is administered intranasally, allowing rapid and efficient brain penetration. It failed in a trial as a monotherapeutic for progressive supranuclear palsy (PSP),[18] a Parkinson's-like disease. However, emulating the vast majority of clinical trials for therapies for neurodegenerative diseases, davunetide was prescribed in a mechanistically naive way: No attempt was made to determine *why* the trial subjects had developed PSP, and no other factors, such as mitochondrial support, were addressed. Despite this failed clinical trial, some doctors have used it as part of an overall treatment protocol for neuroprotection but, again, there haven't yet been many high-quality human studies.

I could go on, but I'm sure you get the point. When it comes to many emerging and increasingly popular peptides, we don't yet know what we don't yet know.

This is not to say that peptide supplementation cannot and will not play an important role in brain maintenance—and soon. Indeed, the pathophysiological model that is at the very heart of the ReCODE Protocol is predicated on the theory that the network insufficiency that characterizes Alzheimer's is associated with an imbalance of two essential peptides that are derived from the processing of the amyloid precursor protein.[19] To me, the question is not *whether* peptide imbalances play a role in cognitive decline—that, in fact, is pretty clear—but *how* to restore the balance in the most physiological, most upstream, safest, and most effective manner.

I know it can be unsatisfying to hear a doctor say, "We'll know more soon," but any doctor who tells you otherwise when it comes to most peptides and cognition, right now, isn't being transparent with you. Guided by a trusted physician, experimentation is a personal choice. But if you're not ready for something like that, you shouldn't be made to feel as though you are missing out. Buy your brain time by doing the many other things that I've discussed in this book. If and when the research catches up with the hype, you'll be in perfectly good cognitive shape to take advantage.

GROWTH FACTORS

Closely related to hormones and peptides are growth factors, also called trophic factors. They are naturally occurring chemicals, typically proteins, that are known to stimulate cellular proliferation, differentiation (for example, from an immature neuron to a mature neuron), and survival. They have, in some cases, been shown to be strongly associated with lifelong brain health.

You've likely heard of the trophic factor most intimately associated with Alzheimer's disease, brain-derived neurotrophic factor (BDNF), which binds to a receptor that comprises two proteins, $p75^{NTR}$ and TrkB. The first of those, $p75^{NTR}$, interacts with the amyloid precursor protein (APP) that is at the center of Alzheimer's disease, and through this mechanism and others it appears that BDNF exerts an anti-Alzheimer's effect.[20] The best-known way to increase BDNF is exercise,[21] and—no surprise—it is the preferable way to increase this growth factor. There are also new drugs being developed that mimic BDNF,[22] and it can be supplementarily increased by whole coffee fruit extract, at 100 to 200 milligrams per day.[23]

Another neurotrophin that plays a central role in Alzheimer's is nerve growth factor (NGF), which supports the very memory-related cholinergic neurons that are heavily affected by this disease. There have been different strategies to increase NGF in Alzheimer's patients, such as transplanting NGF-secreting cells into the brain,[24] but for now the most obvious way to increase NGF is—you guessed it—exercise.[25] There are also a few supplements that increase endogenous NGF, including *Hericium erinaceus* (lion's mane mushroom) at up to 3 grams per day, and acetyl-L-carnitine (ALCAR) at 500 to 1,000 milligrams three times per day.

The benefits are not always linear, though. One of the best-known growth factors, for instance, is a class of signaling molecules we've discussed many times in this book, cytokines, which are vital messengers that help the body mount an all-systems immune response in times of need, but which can also trigger runaway and chronic inflammation and can ignite an inferno of other diseases.[26]

Most people who understand what "bad" growth factors, like some cytokines, can do to our brains would be very wary if someone suggested they inject *more* of these proteins into their bodies. Yet there seems to be quite a bit less hesitance around several growth factors that have, for whatever reason, gotten the reputation for being "good" growth factors. Unanswered questions abound when it comes to what represents a "healthy" balance of these other factors at various points in a very long human life.

We know, for instance, that the insulin-like growth factor, IGF-1, is important to a healthy brainspan. Moreover, low IGF-1 is associated with an increased risk for Alzheimer's and brain atrophy.[27] However, high IGF-1 is associated with cancer. It's unclear what levels are most beneficial at what points in life. Low levels of IGF-1 increase the risk of death for younger adults but decrease all-cause mortality for older people.[28] Interestingly, people in the middle of this pleiotropic mystery—those in their mid-forties to early-sixties—are at an increased risk of dementia if they have *either* very low or very high levels of IGF-1,[29] which may help explain why a team of researchers that included luminary aging researchers Nir Barzilai and Sofiya Milman, both of New York's Albert Einstein College of Medicine, once called this growth factor "the Jekyll and Hyde of the aging brain."[30] Balance is clearly key, and we don't yet know what that balance looks like in any one individual.

Like many of the soonist interventions we're discussing in this chapter, however, this doesn't mean that a physician who observes that your IGF-1 levels are very high or very low, and wishes to intervene, is doing anything wrong. Much to the contrary, if in most other respects a person looks "normal" but this growth factor is disconcertingly outside of a healthy reference range, it might indeed be the right strategy to attack that problem with supplementation. But like the other hormone levels we have spoken of, IGF-1 levels are also malleable with lifestyle interventions. Low protein intake is generally associated with low IGF-1 while high-protein diets are broadly known to increase IGF-1 levels, although these associations have also been demonstrated to be highly dependent on a person's state of biological aging and other lifestyle factors.[31] Zinc, human growth factor, and sleep can all lead to increased IGF-1. Thus, I would be suspicious of a physician who glances at a chart, sees a low level of IGF-1, and immediately calls for supplementation. That hesitancy would be greatly alleviated, however, if the doctor was also deeply engaged with the patient about matters of exercise, sleep, stress, infections, toxins, genetics, and family history of disease.

The same would be true in the case of another growth factor that had a well-known but also very complex association with lifelong brain health: glial cell line–derived neurotrophic factor (GDNF). You might remember that glial cells are an essential part of the glymphatic system, the relatively newly discovered mechanism for waste clearance from our brains and other parts of the central nervous system that we discussed in chapters 8 and 9. A circulating protein derived from these cells has been demonstrated to promote the survival of neurons, and implicated as a protective factor against cognitive decline in animal studies.[32] Human research is furthest along in the relationships between GDNF and Parkinson's disease, but even in that arena the clinical data are limited. It will not shock me at all to find, in five or ten years' time, that researchers have identified ways to best assess healthy levels of GDNF as well as the most effective delivery mechanisms, since it can be tricky to get these chemicals to the parts of the body where they might be most useful. (One strategy that has been tested is the implantation of a device into the brain that slowly releases GDNF over time.[33] This is an interesting idea for people with a rapidly moving, progressively neurodegenerative disease, but it's obviously not a scalable solution for preventative maintenance of GDNF.)

Another protein worth noting for its potential as a soonist intervention for cognitive aging and disease is basic fibroblast growth factor (bFGF). While initially named because of its clear role in the proliferation of fibroblasts, the cells that do the lion's share of the work in forming connective tissue, over the past few decades bFGF has been shown to have a hand in many cellular processes, including the protection of the blood-brain barrier after organ dysfunction or brain trauma.[34] Put another way, this growth factor seems to help heal the structure that protects our brains from a litany of other insults, including toxins and infections. But just like IGF-1 and GDNF, the "right time, right place" qualities of bFGF have not yet been sorted out. I do believe we'll begin meaningfully solving these mysteries, and soon.

Reelin is another protein that is not typically classified as a growth factor. It is thought to regulate the growth of some types of immune cells, play a role in proliferation of other cells, possibly promote wound healing, and regulate cellular differentiation—thus fulfilling many of the same roles as traditional growth factors. Imagine trying to make your way through a thick jungle—it would really be helpful to have someone with a machete there to hack through the growth, right? Well, that is what reelin does—it cuts proteins that hold cells together (cell adhesion molecules), so that brain cells, or their fingerlike processes, can make their way to their targets.

In 2011, it was shown that reelin supplementation enhances cognitive ability and synaptic plasticity in mice.[35] In 2023, many people first heard about reelin following the publication of a case study of a Colombian man with a rare early-onset mutation that should have given him full-blown dementia by his early forties. Instead, he managed to evade the disease for two more decades, apparently because he had another mutation that produced plenty of reelin.[36] That single-patient study was certainly fascinating, and potentially very important, because the protein had previously been implicated in multiple processes including brain development, psychiatric illness, and synaptic plasticity, but by itself it didn't tell us much. Soon, though, researchers published an analysis of four hundred people who retained strong cognitive function as they aged—many of whom had more reelin in their brains.[37] And just a few months after that, another study offered more evidence in an analysis of the brains of several people who had lots of amyloid plaques, which are of course associated with Alzheimer's, and yet no signs of dementia. What those brains did have was lots of reelin.[38]

Not surprisingly, there are ongoing efforts to develop a drug that stimulates reelin production, and there are even discussions about a reelin supplement. But let's be clear: Reelin is a protein, so it would not be effective taken orally. It would have to be injected (as it was in the mouse studies); and there is very little evidence thus far on whether an artificial boost in reelin would have the same effect in humans as it does in mice. Nonetheless, it is an excellent candidate—but let's look at why: When the brain switches modes from connection to protection, there is an elegantly organized set of switches: from making synapses (synaptoblastic) to pulling back (synaptoclastic) and protecting the brain; from anti-inflammatory to proinflammatory; from antithrombotic to prothrombotic; from neurotrophic to antimicrobial; and from stabilizing tau to prionic phospho-tau. So, just as you would predict, part of this programmatic switch is from neuroplasticity (high reelin) to retreat (low reelin). Thus, the programmatic switch is about far more than reelin, even though reelin plays its role in neuroplasticity. Once again, we want to determine what is driving this mode switch. Is it exposure to a new pathogen or toxin, reduced energetics, or something else?

STEM CELLS

From the time that stem cells were theoretically conceived in the 1900s to their actual discovery in the 1960s, the potential of such a cell in fighting disease was very clear. If stem cells were capable of making other cells like themselves, and also many other types of cells in the body, they could likely be used to generate healthy cells to replace diseased cells.[39] If so, we might effectively end virtually any disease! Decades of research later, we're still working on turning that theory into reality, in part because the use of embryonic stem cells—those that are most versatile because they represent the least differentiated state of cellular existence—has been ethically controversial. Over the past twenty years, though, the advent of induced pluripotent stem cells—adult cells that have been pushed into a less differentiated state through genetic manipulation—has helped turbocharge the field of stem cell research, and experimental uses abound for patients in need of bone marrow transplantation, cancer treatment, and therapies for blood and immune system disorders. But in recent years many "rogue clinics" have sprung up around the world offering stem cell therapy, especially those that promise to fix damaged joints but increasingly for other conditions, including for cognitive decline and neurodegenerative diseases. And since would-be patients are sometimes unable to find doctors willing to help them access stem cells in their own countries, a thriving medical tourism industry has emerged, with relatively inexpensive access to unproven therapies of all sorts in other nations,[40] with results that run the gamut from seemingly miraculous healing to life-threatening and sometimes even life-ending complications that include fast-spreading infections, cellular rejection, and the rapid accumulation of tumorous cells.[41]

Stem cells already have widespread application in aging and age-related diseases. For example, there are ongoing trials of stem cells for Alzheimer's disease.[42] However, as you know, Alzheimer's disease is a dynamic, ongoing process, so the idea of treating with nothing but stem cells is a bit like trying to rebuild a house as it's burning down. Wouldn't you want to extinguish the fire first? Therefore, I look forward to stem cell trials that include modern evaluation and personalized treatment, rather than simple administration as a monotherapy. Stem cell–based treatments should soon be a standard method for treating neurological diseases—and likely a preventative measure as well. The research is moving fast, after all.

In just a period of a few months in late 2023 and early 2024, for instance, researchers at the University of California San Diego demonstrated that stem cell transplants were effective in preserving memory and cognition, reducing neuroinflammation, and significantly lessening amyloid accumulation in mice with a rodent version of Alzheimer's disease ("Mouzheimer's").[43] Then, an international team of scientists showed that stem cell therapies may protect people with progressive multiple sclerosis from further brain damage,[44] and shortly thereafter, a clinical trial at Hadassah University Hospital in Israel reportedly demonstrated that repeated treatment with the stem cells led to gains in mobility and cognition for patients with that same disease.[45] And a few weeks later, in the first known human trial of its kind, a team in California injected stem cells directly into the brain of a patient with mild-to-moderate Alzheimer's disease.[46]

With multiple stem cell types, it will be important to determine which are most effective for brain aging and for each neurodegenerative disease. Mesenchymal stem cells (MSCs) are the most commonly used and can be derived from the recipient as autologous transplants, meaning they come from you and go to you. Thus, there are no concerns about exposure to an exogenous virus or other pathogen. MSCs are usually derived from bone marrow, and they naturally target areas of damage, improve immune function, reduce inflammation, support growth of new blood vessels, and help with cellular survival, among other effects.

Adipose-derived regenerative cells (ADRCs) represent another method to receive autologous stem cells, since they are derived from your own body fat. These also exert striking effects on immune regulation and inflammation, trophic factors, tissue repair, and cell survival. MSCs and ADRCs are readily available today.

Another approach is to use stem cells not from yourself but allografts from another person, for example from the umbilical cord, or from Wharton's jelly (between the placenta and umbilical vessels). These are also MSCs but, since they are derived from a younger individual, they may have more potential for growth and division. These MSCs are also available today and may be even more available in the future as autologous transplants, as more people are opting to have their children's umbilical cords frozen for future potential use.

Yet another method is to use induced pluripotent stem cells (iPSCs), in which, typically, skin cells or blood cells are programmed genetically to return to the stem cell state. To date, these have been used mainly to make models of disease states—for example, taking skin cells from a patient with Alzheimer's, engineering them back to iPSCs, and then differentiating them into neurons to study the disease. In addition, iPSCs hold great promise for treatment of degenerative diseases, immune-mediated diseases, cancers, other diseases, and aging itself.

It's hard not to be excited about all this, even if we cannot yet be sure where it's all headed. And I can very much understand why people who are suffering from acute and progressive neurodegeneration might feel they have no choice but to take a chance on whatever treatment they can access right now. I will not fault desperate people for taking desperate measures.

For those who have not been so afflicted, I believe quite strongly that the right move is to work through all the other measures to slow down and reverse cognitive aging and disease, thus buying time for a world in which these therapies have had more time to mature.

REPROGRAMMING

Jae-Hyun Yang may not yet be a household name, but I think he might be someday soon.

In 2020, the Harvard Medical School researcher played a key role in the discovery of a process of genetic manipulation by which vision could be restored in mice with the rodent equivalent of glaucoma. This was a breakthrough study that strongly suggested biological aging might be a result of changes to the epigenome, the chemical compounds that are largely responsible for the various ways in which our DNA is read and expressed, and that a youthful pattern of these chemical markers could be restored.[47] Three years later, Yang was the first author of a paper in the journal *Cell* that demonstrated the same process that had restored vision in those mice might also reverse symptoms of biological aging throughout the body, including substantial improvements in memory.[48] Months later, in the journal *Aging*, Yang and his collaborators showed that a youthful genome-wide reversal of the markers of aging could be quite easily induced through certain chemicals, without the need for genetic manipulation.[49]

There is a lot of understandable excitement around this idea of "reprogramming," and I share that thrill. Of course, it is not the same thing to change the biological trajectories and physiological capacities of mice as it is to do the same thing in humans. If it's possible at all to reverse aging in humans through reprogramming, I think we're likely a few decades away from that reality.

But one of the reasons I'm excited by this isn't because I am confident that reprogramming will be in wide use by humans in a soonist future. Rather, it's because of the speed at which our world moved from a novel idea about how biological aging occurs to a new reality in which researchers have provided substantial evidence that this cause was not inevitable—and might even be rather easily reversible! What this tells me is that there are other ideas about improving the trajectories for brain aging yet to be conceived, and might completely shift our perspectives on what is possible in just a few short years.

When we envision the future, we really have no choice but to extrapolate what we know from the past and the present. But while it is very hard to predict extremely momentous occurrences with specificity, we can know with great certainty that events like these will occur. These are often called black swan events and are generally considered to be negative, but I would argue that science is far more replete with white swans—instances in which an idea seemingly comes out of nowhere and quickly transforms what we conceive to be possible. These are most certainly part of a soonist view of the future. Thus, every day you live with a youthfully functioning brain is a day closer to the next white swan event for cognitive health and longevity.

So, I will say once again: Do what you can now. Keep working toward the goal of a hundred-year brainspan and do not stop. There is so much you can already do, and so much more than was available just a few years ago! And there is more help on the way—some of it quite easy to see on the horizon, some of it yet unimaginable.

It's coming. And it's coming soon.

14

PRESCRIPTION FOR AN AGELESS BRAIN

Youth is no longer wasted on the young; it is increasingly offered to all.

—R. F. LOEB

We can all agree on one thing: Life isn't much fun without the benefit of an optimally functioning brain. So before we part, I wanted to summarize the prescription for an ageless brain: the tests to obtain, the basics we can all do, the specifics for those in need, and the experimental to consider when the time is right.

When should we get these tests? At thirty-five, forty, forty-five, fifty, fifty-five, sixty, and then every two years thereafter. This will guarantee that you'll be able to see things coming and respond effectively.

Next, determine your brain's biological age, as well as your body's biological age. The best way to do this currently is with a DNA methylation test, such as the ones offered by TruDiagnostic. (If you're interested, you can also determine the ages of your other organs, such as the liver and kidney, but at least check your brain and body ages.) Now, the goal is to reach ten years younger than your chronological age in your forties and fifties, fifteen years younger than your chronological age in your sixties and seventies, and twenty years younger in your eighties and nineties. This means that your brain should function quite well for your entire lifespan. The great news, as you know, is that you can reduce your biological age, as has now been demonstrated in several studies.

Next, we want to know *if* we are showing even the earliest changes that may lead one day to a neurodegenerative process, and if so, *why*, so that we can address the underlying contributors. For the *if* testing, there are new, excellent blood tests available that allow virtually everyone to avoid an invasive spinal tap and an expensive PET scan with radioactive tracers:

- P-tau 217 reflects the brain's ongoing biochemical changes associated with Alzheimer's disease, and a high-sensitivity p-tau 217 (which utilizes a special machine from ALZpath and technology called Simoa) is available, allowing all of us to identify the earliest changes years before any symptoms begin, as well as providing a test to follow our progress during treatment. This gives us a huge advantage, allowing virtually everyone to begin earliest treatment and avoid ever progressing through SCI to MCI and finally to dementia. This test should end up saving millions who would otherwise have gone undiagnosed until the relatively late stages of disease, but it will require all of us to reframe what we think about the term *Alzheimer's*. Many people are, understandably, so stressed by hearing this term associated with their own brains that they do not stop to think that this knowledge will help to avoid all Alzheimer's-related cognitive decline. Therefore, it is a key part of ensuring a lifetime of optimal cognition.
- GFAP (glial fibrillary acidic protein) is produced by the cells that are the caretakers of the neurons—the astrocytes (also called astroglia)—and so this tells us whether we have ongoing inflammation and repair in our brains. Unlike p-tau 217, GFAP is not specific for Alzheimer's, but it is very sensitive, so

if we have ongoing insults such as chronic infections, GFAP will increase. It also increases years before neurodegenerative diseases become symptomatic, so it is a wonderful early-warning system and complementary to p-tau 217.
- NfL (neurofilament light) is complementary to p-tau 217 and GFAP in that it indicates neuronal damage from any source, such as traumatic brain injury, vascular disease, or neurodegeneration.
- Beta amyloid 42:40 ratio is an Alzheimer's-specific marker that declines as Alzheimer's is developing.
- Syn-One is a skin biopsy test that indicates the presence of three different diseases: Parkinson's, Lewy body disease, and MSA (multiple system atrophy). Parkinson's and Lewy body disease are very common; there are over a million Americans with Parkinson's and over a million with Lewy body disease. Again, early warning is very beneficial and will allow you to avoid severe symptoms.

These tests are summarized in Table 1.

Now, if you find out that you have an abnormal test, you'll want to find out why it is abnormal. Similarly, if your tests are normal, you'll want to know why you still may be at risk, and these tests are summarized in Table 2. Please do not worry—getting on an optimal, personalized protocol should help you to return the test values to normal and avoid decline.

The *why* tests will identify the potential contributors to brain aging and degeneration, ensuring that you'll have the best chance of life-long cognitive health. As you no doubt already know, any risk may be associated with both genetics and environment, thus we want to evaluate both genetics and biochemistry. As Lee Hood noted, it's about genome and phenome.

Once you have these tests, if you find unexplained results, there are follow-up tests (also in Table 2) to help explain—the basis of twenty-first-century medicine is to keep asking *why* until everything makes sense. We do not develop disease for no reason—that's twentieth-century medicine, to diagnose and prescribe. Human pathophysiology is complicated, but understanding why you are at risk for brain aging and degeneration (or why you are experiencing such changes) makes all the difference, offering outcomes never before achieved.

Instead of a generic approach that may be helpful for some but not all, we can adopt a personalized approach, which consists of three parts: basics (Table 3), specifics (Table 4), and experimentals to consider (Table 4).

Table 1. Noninvasive tests for early detection of neurodegeneration

TEST	RATIONALE	TARGET	COMMENT
p-tau 217	Early marker for Alzheimer's; allows follow-up over time	Depends on lab; <0.34 ng/L for Neurocode	ALZpath machine and Simoa technology; currently the most sensitive
GFAP	Early marker for inflammation and repair	0–120 ng/L	Produced by activated astroglia; not specific for AD
NfL	Marker of neuronal damage	Age-dependent*	May be increased with trauma, vascular disease, or any cause of neuronal damage
Beta amyloid 42:40 ratio	Marker for brain amyloid	>0.170:1 (Quest)	Reduced in AD
Syn-One (from skin biopsy)	Marker of Parkinson's, Lewy body, or MSA	Negative	Identifies alpha-synuclein aggregates

* Age-related normal values for NfL, by age group:

20–29 (≤ 8.4 ng/L)   50–59 (≤ 20.8 ng/L)   ≥ 80 (≤ 51.2 ng/L)
30–39 (≤ 11.4 ng/L)   60–69 (≤ 28.0 ng/L)
40–49 (≤ 15.4 ng/L)   70–79 (≤ 37.9 ng/L)

Table 2. Target values of some biochemical and physiological tests associated with cognition (adapted from *The End of Alzheimer's Program*)

	CRITICAL TESTS	TARGET VALUES	COMMENTS
INFLAMMATION, PROTECTION, AND VASCULAR	hs-CRP	<0.9 mg/dL	Systemic inflammation
	Fasting insulin Fasting glucose Hemoglobin A1c HOMA-IR	3.0–5.5 μIU/mL 70–90 mg/dL 4.0–5.3% <1.3	Glycotoxicity and insulin resistance markers
	Body mass index (BMI)	18.5–25	Weight (lbs) × 703/height (inches)^2
	Waist to hip ratio (women) Waist to hip ratio (men)	<0.85 <0.9	
	Homocysteine	≤7 μmol/L	Reflects methylation, inflammation, and detox
	Vitamin A (retinol)	38–98 mcg/dL	May be toxic at higher levels
	Vitamin B$_6$ Vitamin B$_9$ (folate) Vitamin B$_{12}$	25–50 mcg/L (PP) 10–25 ng/mL 500–1500 pg/mL	All improve methylation and reduce homocysteine
	Vitamin C Vitamin D Vitamin E	1.3–2.5 mg/dL 50–80 ng/mL 12–20 mg/L	

	CRITICAL TESTS	TARGET VALUES	COMMENTS
	Omega-6 to omega-3 ratio	1:1 to 4:1 (beware that <0.5:1 may be associated with bleeding tendency)	Ratio of inflammatory to anti-inflammatory omega fats
	Omega-3 index	≥10% (ApoE4+) 8–10% (ApoE4-)	Proportion of anti-inflammatory omega-3 fats
	AA to EPA ratio (arachidonic acid to eicosapentaenoic acid ratio)	<3:1	Ratio of inflammatory AA to anti-inflammatory EPA
	A/G ratio (albumin to globulin ratio) Albumin	≥1.8:1 4.5–5.4 g	Markers of inflammation, liver health, and amyloid clearance
	LDL-P Small dense LDL Oxidized LDL	700–1200 nM <28 mg/dL <60 ng/mL	LDL-P is LDL particle number
	Total cholesterol HDL cholesterol Triglycerides TG to HDL ratio	150–225 mg/dL 50–100 mg/dL 40–90 mg/dL <1.3	
	ApoB	40–90 mg/dL	Measures multiple atherogenic particles
	CoQ10	1.1–2.2 mcg/mL	Affected by cholesterol level
	Glutathione	>250 mcg/mL (>814 μM)	Major antioxidant and detoxicant

	CRITICAL TESTS	TARGET VALUES	COMMENTS
	Leaky gut, leaky blood-brain barrier, gluten sensitivity, autoantibodies	Negative	
MINERALS	RBC-magnesium	5.2–6.5 mg/dL	Preferable to serum magnesium
	Copper Zinc	90–110 mcg/dL 90–110 mcg/dL	
	Selenium	110–150 ng/mL	
	Potassium	4.5–5.5 mEq/L	
TROPHIC SUPPORT	Vitamin D	50–80 ng/mL	(25 hydroxy vitamin D3)
	Estradiol Progesterone	80–250 pg/mL 1–20 ng/dL (P)	Women; age dependent
	Pregnenolone Cortisol (AM) DHEA-S (women) DHEA-S (men)	100–250 ng/dL 10–18 mcg/dL 100–380 mcg/dL 150–500 mcg/dL	Age dependent
	Testosterone Free testosterone	500–1000 ng/dL 18–26 pg/mL	Men; age dependent

	CRITICAL TESTS	TARGET VALUES	COMMENTS
	Free T3 Free T4 Reverse T3 TSH Free T3 to reverse T3 Antithyroglobulin antibodies Anti-TPO	3.2–4.2 pg/mL 1.3–1.8 ng/dL <20 ng/dL <2.0 mIU/L >0.02:1 Negative Negative	Milli-international unit per liter (mIU/L) = micro-international unit per milliliter (μIU/mL)
	BDNF	>90 pg/mL	Brain-derived neurotrophic factor
TOXIN-RELATED	Mercury Lead Arsenic Cadmium	<5 mcg/L <2 mcg/dL <7 mcg/L <2.5 mcg/dL	Heavy metals (provoked and urinary tests also used)
	Mercury Tri-Test	<50th percentile	Hair, blood, urine
	Organic toxins (urine)	Negative	Benzene, toluene, etc.
	Glyphosate (urine)	<1.0 mcg/g creatinine	Herbicide
	Copper to zinc ratio	0.8–1.2:1	Higher ratios associated with dementia
	C4a TGF beta 1 MMP-9 MSH	<2830 ng/mL <2380 pg/mL 85–332 ng/mL 35–81 pg/mL	Associated with inflammatory response

	CRITICAL TESTS	TARGET VALUES	COMMENTS
	Urinary mycotoxins	Negative	May include contributions from inhalation, ingestion, and infection
	BUN Creatinine	<20 mg/dL <1.0 mg/dL	Reflects kidney function
	AST ALT	<25 U/L <25 U/L	Reflects liver damage
	VCS (visual contrast sensitivity)	Pass	Failure associated with biotoxin exposure
	ERMI test	<2	Mold index from building
	HERTSMI-2 test	<11	Index of most toxic molds
PATHOGEN-RELATED	CD57	60–360 cells/uL	Reduced with Lyme disease
	MARCoNS	Negative	Multiple antibiotic resistant coagulase-negative *Staphylococci*
	Antibodies to tick-borne pathogens and FISH test	Negative	*Borrelia, Babesia, Bartonella, Ehrlichia, Anaplasma*
	Antibodies to *Herpes* family viruses	Negative	HSV-1, HSV-2, HHV-6A, EBV, CMV

	CRITICAL TESTS	TARGET VALUES	COMMENTS
	Antibodies to *Chlamydia pneumoniae*	Negative	
NEURO-PHYSIOLOGY	Peak alpha frequency on quantitative EEG	8.9–11 Hz	Slows with cognitive decline; useful for following progress
	P300b on evoked response testing	<450 ms	Delayed with cognitive decline; useful for following progress
IMAGING	MRI with volumetrics	Negative for atrophy, vascular disease, white matter disease, hydrocephalus, neoplasm, and other pathology	ASL sensitive for Alzheimer's
	Cone beam CT	Negative for oral abscesses	Important in those with root canals
	PET scan (amyloid tracer)	Negative for amyloid	
	PET scan (FDG)	Normal glucose metabolism	Reduced in temporal and parietal lobes in AD
	PET scan (F-DOPA)	Normal DOPA uptake	Reduced in Parkinson's

	CRITICAL TESTS	TARGET VALUES	COMMENTS
OTHER TESTS	MoCA (Montreal Cognitive Assessment) or SLUMS (St. Louis University Mental Status)	28–30	
	Nocturnal oxygen saturation (SpO$_2$)	96–98%	Affected by living at high altitude
	Sleep study (or data via wearable)	Total sleep 7–9 h; REM ≥1.5 h; deep ≥1 h	
	AHI (apnea/hypopnea index)	<5	>5 indicates sleep apnea
	Oral DNA	Negative for pathogens	*P. gingivalis*, *T. denticola*, etc.
	Stool analysis	No pathogens or dysbiosis	
	ImmuKnow (CD4 function, indicated by ATP production)	≥525 ng/mL	Indicates function of helper cells of the cellular arm of the adaptive immune system
	Epigenetic testing (e.g., TruDiagnostic)	Biological age and brain age < chronological age	

	CRITICAL TESTS	TARGET VALUES	COMMENTS
	Genetics (e.g., IntellxxDNA, 3x4 Genetics, DNA Life)	Negative for ApoE4, TREM2 mutations, LRRK2 mutations, and other mutations associated with neurodegeneration; reduced detox, inflammation, protein turnover defects, and hypercoagulability. Heterozygous for klotho VS	Whole-genome sequencing available from Nebula, HLI, and Sequencing

Abbreviations: AA, arachidonic acid; AD, Alzheimer's disease; AHI, apnea/hypopnea index; ALT, alanine aminotransferase; ASL, arterial spin labeling; AST, aspartate aminotransferase; BMI, body mass index; BUN, blood urea nitrogen; C4a, complement split product 4a; CD57, cluster of differentiation 57; CoQ10, coenzyme Q10 (ubiquinone); DHEA-S, dehydroepiandrosterone sulfate; DNA, deoxyribonucleic acid; DOPA, dihydroxyphenylalanine; EEG, electroencephalogram; EPA, eicosapentaenoic acid; ERMI, Environmental Protection Agency relative mold index; FISH, fluorescent in situ hybridization; HERTSMI-2, Health Effects Roster of Type-Specific Formers of Mycotoxins and Inflammagens—2nd Version; HOMA-IR, homeostatic model assessment of insulin resistance; hs-CRP, high-sensitivity C-reactive protein; LDL, low-density lipoprotein; MARCoNS, multiple antibiotic-resistant coagulase negative *Staphylococcus*; MMP-9, matrix metalloprotease-9; MoCA, Montreal cognitive assessment; MSH, alpha-melanocyte stimulating hormone; P300b, positive wave at 300 milliseconds (event-related potential), component B; PP, pyridoxal phosphate; RBC, red blood cell; SpO_2, peripheral capillary oxygen saturation; T3, triiodothyronine; T4, thyroxine; TG, triglycerides; TGF beta 1, transforming growth factor beta-1; TPO, thyroid peroxidase; TSH, thyroid-stimulating hormone.

Table 3. The seven basic modalities for a youthful brain

MODALITY	GOAL	RATIONALE	COMMENT
DIET	Plant-rich, mildly ketogenic diet	Create metabolic flexibility, optimize energetic support, minimize inflammation and toxicity	Metabolic syndrome commonly reverses; many no longer require antihypertensives, statins, or anti-diabetes drugs
EXERCISE	Combine aerobics and strength training	Multiple benefits: blood flow, mitochondrial function, insulin sensitivity, improved sleep, etc.	Consider Kaatsu for muscle mass, EWOT for blood flow and oxygenation, cold water for mitochondria
SLEEP	7–8 h; ≥ 1 h deep; ≥1.5 h REM; SpO_2 >94%	Multiple benefits: cognition, detox, repair, immune system, etc.	Deep sleep key for detox; REM and non-REM both key for memory
STRESS MANAGEMENT	HRV optimal for age (>70 at 35, >65 at 55, >60 at 65+)	Multiple benefits re cortisol, glucose, blood flow, etc.	Improvements in limbic system, vagal tone, immune system, etc.
BRAIN STIMULATION	Continued improvement; >75th percentile for age	Stimulate synapse formation and maintenance	Many methods: brain training, PBM, MeRT, microcurrent, etc.
DETOXIFICATION	<50th percentile for inorganics, organics, and biotoxins	Reduce ongoing compromise of energetics, inflammatory effects	Many methods, including sauna, crucifers, binding agents like charcoal and bentonite clay

MODALITY	GOAL	RATIONALE	COMMENT
TARGETED SUPPLEMENTS	Optimal levels of nutrients, hormones, trophic factors, etc.	Avoid deficiencies, support optimal biochemistry	See Table 4

Table 4. Some examples of the specific therapeutics for a youthful brain

INTERVENTION TYPE	TREATMENT	RATIONALE	COMMENTS
SUPPLEMENTS—VITAMINS	Vitamin A	Supports neuroplasticity, reduces risk for macular degeneration	Typical dose is 900 mcg/d
	Vitamin B_1 (thiamine)	Supports memory and energy	Typical dose 5–100 mg (as benfotiamine, 150–600 mg/d)
	Biotin (B_7)	Reduces depression, improves skin	50–100 mcg/d is typical
	Methyl-B_{12}, methylfolate, P5P	Reduces homocysteine	Some people with hypermethylation (e.g., COMT mutations) will require nonmethylated forms.
	Vitamin C	Antioxidant for aqueous environments (nonmembranes)	Typical dose 500–1000 mg/d

INTERVENTION TYPE	TREATMENT	RATIONALE	COMMENTS
	Vitamin D and K2	Multiple effects of D—brain health, tumor inhibition, bone support, cardiovascular, etc.	100 mcg of K2 should be included for any vitamin D dose >1000 IU
	Vitamin E	Antioxidant for lipid environments (membranes)	400–800 IU as mixed tocopherols and tocotrienols
SUPPLEMENTS— MINERALS	Magnesium (as citrate or glycinate or threonate or chelate)	Multiple mechanisms: improves gut motility, calms nervous system, improves insulin sensitivity, etc.	Mg citrate helpful for constipation; Mg threonate for brain penetration. Typical dose 500 mg/d (for Mg threonate, 150 mg of Mg/d)
	Zinc (as picolinate or other)	Multiple mechanisms: increases insulin secretion, supports immune system, etc.	Typical dose 15–50 mg
	Iodine	Thyroid support	Kelp and seaweed are good sources; RDA is 150 mcg/d
	Copper	Supports multiple enzymes	Zinc and copper compete, so beware of overdoing either
	Potassium	Reduces risk for atrial fibrillation; supports muscles and nerves	Low potassium associated with risk for dementia

INTERVENTION TYPE	TREATMENT	RATIONALE	COMMENTS
	Calcium	Bone support	Total intake should be 2–2.5 g/d, most of which comes from diet
	Iron	Hemoglobin for carrying oxygen; multiple enzymes	Toxic at high doses, especially combined with vitamin C
SUPPLEMENTS— HERBS AND RELATED	Curcumin	Anti-inflammatory with strong binding to amyloid and tau	Typical dose of 0.5–1.5 g/d, but some use as high as 10 g/d
	Ashwagandha	Increases amyloid removal; calms	May interfere with thyroid
	Bacopa monnieri	Increases acetylcholine for memory	Typical dose 250–500 mg with meals
	Rhodiola	Adrenal support	Typical dose 200–600 mg/d
	Gotu kola	Enhances focus and concentration	Typical dose 500 mg twice per day
	Shankhpushpi	Memory and concentration support	Typical dose 250–500 mg/d
	Mucuna pruriens	Increases dopamine	Often used for Parkinson's disease

INTERVENTION TYPE	TREATMENT	RATIONALE	COMMENTS
	Triphala (Amalaki, Haritaki, and Bibhitaki)	Supports gut health	
	Guduchi (*Tinospora cordifolia*)	Immune support	Often helpful for fungal infections
	Ginkgo biloba	Vascular support	Failed in trial as AD monotherapy
ENERGY AND STRUCTURE	Creatine	Increases energy	Usually as monohydrate; typical dose 5 g/d
	Nicotinamide riboside (NR), nicotinamide mononucleotide (NMN), or niacinamide	Increase NAD+/energy	Some also use NAD+ IV, which is particularly helpful for those with cognitive decline after chemotherapy ("chemobrain")
	Ketone salts, esters, or mixtures	Increases ketone level	Often discontinued when endogenous ketosis is achieved
	MCT oil, coconut oil	Increases ketone level	May exacerbate hyperlipidemia
	Collagen	Provides amino acids for protein synthesis	

INTERVENTION TYPE	TREATMENT	RATIONALE	COMMENTS
	Whey protein	Provides amino acids for protein and muscle building	
	Glycine	Supports protein synthesis	Typical dose 5 g/d
	Urolithin A	Supports mitochondrial renewal and energy	Typical dose 250–1000 mg/d
	Methylene blue	Supports mitochondrial electron flow	Promising for Parkinson's, possibly Alzheimer's
	EWOT, HBOT	Increase blood flow and oxygenation	Especially helpful for those with vascular disease or traumatic brain injury
SUPPLEMENTS—DETOX	S-acetyl glutathione or liposomal glutathione	Increase glutathione for detox and antioxidant effects	Consider IV glutathione or intranasal for those with significant toxicity
	N-acetylcysteine (NAC)	Precursor to glutathione	Typical dose 500–600 mg once or twice per day
	Sulforaphane	Increase detoxifying glutathione	Typical dose is 400 mcg/d
	Organic psyllium husk, konjac root, or other fiber	Supports gut microbiome, increases detox	Target total dietary fiber to >30 g/d

INTERVENTION TYPE	TREATMENT	RATIONALE	COMMENTS
	Chlorella	Binds and helps remove toxins	Often used for metal toxicity
	Binders (cholestyramine, Welchol, activated charcoal, bentonite clay, etc.)	Bind and help remove various toxins	Note that too-rapid detox may be associated with exacerbation of cognitive decline
	Nrf2 activators	Activate endogenous detox pathways	Used cyclically
SUPPLEMENTS— GUT HEALTH	Probiotics, prebiotics, fermented vegetables	Enhance gut microbiome	Gut microbiome diversity associated with improved health and cognition
	Saccharomyces boulardi	Enhances gut microbiome, treats dysbiosis and SIBO	Often used for fungal dysbiosis
	Bone broth	Heals gut	Kettle & Fire brand and similar, or homemade
	Pro-Butyrate	Supports gut health and gut-brain connection	Postbiotics such as butyrate support cognition
	Atrantil	Treats SIBO (small intestinal bacterial overgrowth)	

INTERVENTION TYPE	TREATMENT	RATIONALE	COMMENTS
GLYCEMIC CONTROL	Berberine	Activates AMPK	Most on plant-rich ketogenic diet will not need additional glycemic control
	Metformin	Activates AMPK	Metformin may increase risk for Alzheimer's, Parkinson's
	Semaglutide, tirzepatide	GLP-1 agonists	Also cause early satiety, weight loss; in trials for Alzheimer's
	Many supplements: cinnamon, chromium, purslane, milk thistle, bitters, glycine, fenugreek, etc.	Various mechanisms	
	Bergamot	AMPK	Often used to lower lipids, as well
VASCULAR OPTIMIZATION	Nattokinase, lumbrokinase, pycnogenol	Reduce blood clotting	High-dose nattokinase was associated with regression of arterial plaque
	Nitric oxide (via precursors such as L-Arg, beet root juice, or drugs such as sildenafil)	Dilates blood vessels, increasing blood flow	Use is associated with reduction in risk for Alzheimer's

INTERVENTION TYPE	TREATMENT	RATIONALE	COMMENTS
	Arterosil hp	Endothelial glycocalyx support	Helps heal blood vessels
	Ginkgo biloba	Increases blood flow	
ORAL CARE	Probiotic toothpaste	May improve oral microbiome	
	Oral rinses (e.g., Stella Life, Dentalcidin)	May improve oral microbiome	
	Water flossing	Helps remove food and prevent plaque	
	Removal of dental amalgams	Reduces mercury	Should be done by a trained dentist (see International Academy of Oral Medicine and Toxicology, or IAOMT)
	Removal of infected root canals or abscesses	Reduces inflammation	Many root canals are ongoing sources of inflammation
INFLAMMATORY, IMMUNE, AND ANTIMICROBIAL	DHA/EPA (omega-3)	Anti-inflammatory	Typical dose 1 g/d

INTERVENTION TYPE	TREATMENT	RATIONALE	COMMENTS
	Krill oil	Anti-inflammatory	Includes omega-3 phospholipids
	Resolvins	Resolve ongoing inflammation	SPM Active or similar
	Modified citrus pectin	Anti-inflammatory via Gal-3	Pectasol brand and others
	Glycine	Multiple mechanisms: anti-inflammatory, amino acid for protein synthesis, glucose control, etc.	Common dose 5 g/d
	Low-dose naltrexone (LDN)	Reduces auto-immunity	Typical dose 1–6 mg at night; note that thyroid medication dose may need to be reduced
	Valacyclovir	Antiviral used for *Herpes* viruses, both outbreaks and suppression	Alternatives include acyclovir, L-lysine, monolaurin, and others
	Zinc	Immune support	15–50 mg is typical dose

INTERVENTION TYPE	TREATMENT	RATIONALE	COMMENTS
	AHCC (active hexose correlated compound), transfer factor PlasMyc, humic acid, fulvic acid, and other antivirals	Antiviral immune support	
	Plasma exchange	Anti-inflammatory, detox	Shown to support cognition
	IVIG	Anti-inflammatory, immune supportive	
	Hydrogen water		Unproven
SUPPLEMENTS— OTHER	Choline (as citicoline or α-GPC)	Increases acetylcholine, key neurotransmitter for memory	Dietary goal is 450–550 mg/d
	Huperzine A	Inhibits breakdown of acetylcholine, thus often improving memory	Typical dose 100–200 mcg
	Whole coffee fruit extract	Increases BDNF	Typical dose 100–200 mg
	ALCAR	Increases NGF	Typical dose 500 mg

INTERVENTION TYPE	TREATMENT	RATIONALE	COMMENTS
	Phosphatidyl serine	Multiple effects	Typical dose 300–800 mg/d
	Homotaurine	Inhibits amyloid-beta oligomerization	Promising, especially for ApoE4 individuals
	Plasmalogens	Increase key lipid membrane components	Markedly reduced in Alzheimer's
	Quercetin, Fisetin	Multiple mechanisms, including senolytic	Destroy senescent cells, improve glucose utilization, support immune system, etc. Typical dose 500 mg (quercetin), 300 mg (fisetin)
	Nootropics (aniracetam, piracetam, and similar)	Multiple mechanisms, especially glutamate receptor activation	Mixed results, with some trial failures
	Adderall	Focus	Long-term use discouraged; helpful for those with ADHD or ADD
BHRT	Estradiol, progesterone, testosterone	Bind to intracellular receptors, affect many genes; support cognition	Target estradiol level of 50–100 pg/mL in postmenopausal women; progesterone 2–20 ng/mL

INTERVENTION TYPE	TREATMENT	RATIONALE	COMMENTS
	Testosterone	Supports cognition, musculature, and other effects	Target 500–1,000 ng/dL (men)
	Pregnenolone	Master steroid, precursor to both sex steroids and stress steroids	Target 100–250 ng/dL; typical dose 10–100 mg/d
	DHEA	Supports stress responses	Target 100–350 mcg/dL; typical dose 25–100 mg/d
	Thyroid	Affects many genes, increases metabolism and cognition	Combinations such as Armour more physiological than thyroxine
	Melatonin	Replaces the age-associated decline	Common dose 1–3 mg each night
SLEEP APNEA TREATMENT	CPAP or BiPAP	Positive pressure avoids airway collapse and hypoxemia	Important to optimize settings for best results. Note that sleep apnea goes undiagnosed in 80% of cases.
	Dental device	Holding jaw forward to avoid airway collapse	Best for those with only mild sleep apnea
	Airway expansion	Increase airway size to prevent collapse	Vivos or similar

INTERVENTION TYPE	TREATMENT	RATIONALE	COMMENTS
	Surgical procedures	Increase airway size	Uvulectomy, etc.
BRAIN STIMULATION	Brain training	Improves cognition, reduces risk for decline	BrainHQ, Elevate, etc.
	Photobiomodulation (PBM)	Studies support improved cognition	Gamma frequency may have best effect
	Magnetic stimulation	Similar effects to PBM	TCMS, MeRT, and related
	Microcurrent	Similar effects to PBM	
	Sound	Still in evaluation	Optimal frequency may be 40 Hz range
STEM CELLS	ADRCs (adipose-derived regenerative cells)	Multiple mechanisms: trophic factors, immune support, etc.	May be helpful for aging and multiple neurodegenerative diseases
	Bone marrow–derived	Similar effects to ADRCs	Autologous
	Cord blood–derived	Similar effects to ADRCs	Allografts

INTERVENTION TYPE	TREATMENT	RATIONALE	COMMENTS
	iPSC (induced pluripotent stem cells)	Theoretically with potential to treat many diseases	Experimental, in development
	Laser-guided stem cells		*Unproven
PEPTIDES	Epitalon, Pinealon, Cerebrolysin, Davunetide, Thymosin alpha 1, Thymosin beta 4, and hundreds of others	Many mechanisms, from trophic effects to anti-inflammatory to regenerative effects, etc.	Great potential, some with excellent results, but FDA is making it increasingly difficult for compounding pharmacies to provide

Throughout my career, so many people have waited to see me until they had severe symptoms and would have done anything to have the opportunity that we now all have—early identification, characterization, and protection, so that the tragedies of accelerated brain aging and neurodegeneration never occur. Let's look into the future and walk through a few examples that illustrate how important it is to stay on top of brain health, beginning in middle age.

Susan was full of life and loved to party, but when she turned thirty-five, she checked her basic brainspan labs and found out that she was ApoE 3/4 (and thus at risk for Alzheimer's), her p-tau 217 was normal, but her GFAP was slightly elevated (indicating some brain inflammation and repair) and her biological age was actually a bit high at thirty-nine. Her HOMA-IR (homeostatic model assessment for insulin resistance) was 2.0 (indicating some insulin resistance), she had borderline hypertension at 139/90, and her ApoB was 105. These indicated that, like over 80 million Americans, she was developing metabolic syndrome, a very common risk factor for early brain aging.

She took a deep breath and resolved not to worry, since she knew that the armamentarium for reducing brain aging is large and growing. She began a plant-rich, mildly ketogenic diet with fourteen to sixteen hours of fasting each night, avoiding grains, dairy, and simple carbs; used an over-the-counter continuous glucose monitor (CGM) to follow her glucose; and checked her ketone levels for a month. She found that she was typically in the 0.5 to 1.5 mM Beta-hydroxybutyrate level. She cycled out of ketosis once or twice each week by adding some sweet potato or increasing her resistant starches.

She increased her exercise to five times per week and included both aerobics and weight training. She used Kaatsu bands to increase strength more readily. She lost fifteen pounds and her blood pressure dropped to 118/74, her HOMA-IR dropped to 1.2, and her ApoB was reduced to 82. After one year, her GFAP had returned to normal, indicating resolution of her mild brain inflammation.

She did well on her forty-year labs, with a biological age of thirty-five, normal blood pressure, no insulin resistance, normal lipids, and normal brain biomarkers. However, at forty-two, forty-three, and again at forty-five, she developed COVID-19 and noted some decline in overall energy. She was stressed at work, and when she checked her sleep status on her wearable, she found that she was typically getting only six hours of sleep each night, with only twenty minutes of deep sleep, sixty minutes of REM, although with a good SpO_2 of 95 percent. Her GFAP had increased once again, and her p-tau 217 was now borderline, with a normal NfL. She addressed her stress by taking some time off, doing some shinrin-yoku, and beginning her mornings with some mindfulness meditation. Her energy began to return, her deep sleep increased to nearly sixty minutes per night, her REM to over ninety minutes, and total sleep increased to just over seven hours. After one year, her GFAP and p-tau had returned to normal, and her biological age sat at forty.

At fifty, she began BHRT (bio-identical hormone replacement therapy) and felt great. She had a calcium scan of her cardiac vessels and received a score of eight, indicating low risk. Her bone density was excellent. However, by fifty-five, she noticed that something had changed: She was having some trouble organizing and planning at work, and when she purchased a new mobile phone, she wasn't as quick as usual to learn its various features. She took a free online cognitive assessment and scored in the normal range but realized she had some SCI (subjective cognitive impairment). Her GFAP was elevated at 150 ng/L, and her p-tau was also up at 0.50 ng/L. She wondered why, at only fifty-five, she was developing these early stages of Alzheimer's, and it didn't take long for her trained neurologist to determine why: Her urinary mycotoxins were high, both for trichothecenes (produced by the toxic black mold *Stachybotrys*) and gliotoxin (produced by *Aspergillus*). The neurologist told her not to worry, because even though Alzheimer's used to be considered untreatable, he had personally seen dozens of patients reverse their cognitive changes, especially those in the early SCI phase. Indeed, she was treated with the detox protocol developed by Ritchie Shoemaker, and she went through the DNRS program (dynamic neural retraining system) of Annie Hopper. Over the next twelve months she noted a return to normal. She was once again organizing and planning as she had always done, and she had no problems with her phone, computer, or other tech. Her p-tau and GFAP had come back almost to normal, and by two years, they were completely normal once again. By the age of sixty, her biological age was measured at forty-nine, and she felt great.

Over the next decade, she continued her plant-rich diet, exercised on most days, added some occasional EWOT (exercise with oxygen therapy), tracked her sleep, did some square breathing and meditation to help manage stress, started some brain training, and had no trouble improving her scores over the months. She added some photobiomodulation with a gamma device, continued basic detox, and took several supplements in addition to her BHRT. These included vitamin D (with

100 mcg vitamin K$_2$) to keep her level in the 50 to 80 ng/mL range, Ashwagandha 500 mg twice per day with meals, DHA/EPA 1g/d, nicotinamide riboside 250 mg/d, ProButyrate 600 mg, urolithin A 500 mg, and zinc picolinate 20 mg. She noted that taking magnesium citrate prevented constipation and helped her sleep quality. She also took melatonin 1 mg at night. She checked her gut and found that she did not have leaky gut but did have some SIBO (small intestinal bacterial overgrowth), which responded well to a few weeks of *Saccharomyces boulardi* and some Atrantil. At seventy, her biological age was measured at fifty-five, and her p-tau and GFAP were in the normal range. Her cognition remained excellent, and she enjoyed pickleball and texting with her family members (especially sharing cute animal videos).

At eighty, her physician noted that her calcium score had drifted up to 15, suggesting some mild plaquing. She added a vascular triad for treatment: nattokinase 4000 units/d, nitric oxide (using a beet-derived supplement), and Arterosil hp. She also did more regular EWOT and noted that her vascular age (measured by elasticity) came down rapidly with EWOT. Her retinal exam showed no suggestion of age-related macular degeneration, and the ophthalmologist explained that her ApoE4 gene variant, while it increased risk for Alzheimer's, actually decreased risk for macular degeneration. Nonetheless, she wore blue blockers and avoided bright lights at night. At eighty-five, her calcium score was back down to 7, and her biological age was sixty-seven.

Into her nineties, she slowed things down a bit, reflected on her life, enjoyed her family and friends, and continued the protocol as she had been doing for decades. Her doctor and she discussed the fact that she had defeated Alzheimer's, avoiding dementia by starting early and continuing to optimize her brain aging and health. She kept her brain energetics high, inflammation and toxicity low, managed her stress, and sought help when any early symptoms appeared. At ninety-nine, a friend asked her how long she would like to live, and she said, "As long as I have something to live for." Amen, Susan.

Her family friend Carlton had resolved to live his maximum lifespan with excellent brain function and had actually started at thirty, with a similar diet to Susan's, excellent sleep as documented by his wearable, exercise five days per week, minimizing stress, and keeping his weight optimal. His blood pressure was 110/70, his BMI 22, his HOMA-IR 1.0, ApoB 65, hs-CRP 0.3, and his p-tau 217, GFAP, and NfL all normal. He kept himself in excellent shape through his thirties, forties, and into his fifties, checking his brainspan labs every five years. His blood pressure remained excellent, he remained insulin sensitive, his ApoB was typically in the 60 to 70 range, his calcium score was 0, and his biological age was typically fifteen years younger than his chronological age.

However, at fifty-seven, something strange happened: His wife told him that he had flung his arms wildly while he was sleeping. He looked at his sleep tracker and realized that this had occurred during a REM cycle, which fit with the fact that he had been dreaming about fighting off a dragon. After this had happened for the third time and his wife had threatened to banish him to the couch so that she wouldn't be bruised, he consulted a neurologist, who told him that he had developed REM behavioral disturbance (RBD), something that affects about three million Americans. Carlton read about RBD and learned that it is a common symptom in people who are on the way to developing Parkinson's disease or Lewy body dementia. He learned that RBD is often accompanied by two other pre-Parkinson's symptoms: constipation and anosmia (loss of the sense of smell). In fact, he had noticed some constipation and his sense of smell had not been great for the past several years.

He therefore underwent a Syn-One skin biopsy, which detects the protein associated with both Parkinson's and Lewy body dementia, called alpha-synuclein. Three areas of skin were biopsied, and two did indeed show the presence of alpha-synuclein. He was surprised and somewhat depressed, and he asked his physician what might be causing this problem and what might be done. His physician said, "What do you mean, what causes it? Parkinson's causes it, and we can give you a drug combination of levodopa and carbidopa when you have significant symptoms. Just wait for it, and meanwhile, don't make any long-term plans."

After hearing that, Carlton fired his doctor and found a practitioner who focused on determining the root causes of chronic illnesses like Parkinson's and Lewy body. His new doctor determined that he had had heavy exposure over the years to pesticides such as paraquat, herbicides such as acrolein and glyphosate, and perhaps most importantly, he was living very close to a Superfund site (no one had even mentioned this to him before), where toxic chemicals had been dumped. To make matters worse, his genetics showed that he was set up for toxin-related diseases, with deletions in some of his genes related to glutathione, a key detoxification peptide. So despite the fact that he was a very healthy guy overall, he was on his way to developing a debilitating neurodegenerative condition, which would accelerate his brain aging and ultimately take his life.

His doctor told him that he had really dodged a bullet by getting evaluated before any of the typical motor symptoms of Parkinson's—such as tremor, rigidity, slowness of movements, or instability—had occurred, and before any of the typical symptoms of Lewy body disease—such as visual hallucinations, delusions, and dementia—had occurred. His doctor also checked his gut microbiome, but unlike many, he did not turn out to have the *Desulfovibrio* bacteria often associated with Parkinson's.

Carlton vowed to address the various contributors, and he moved away from the Superfund site, increased his glutathione (initially with IV glutathione, and then with oral S-acetyl glutathione 300 mg), made sure he ate organic fruits and vegetables, took PQQ 20 mg to increase mitochondrial number, urolithin A 500 mg to enhance mitochondrial turnover and function, coenzyme Q 500 mg, and continued his exercise routine, adding some coordination exercises. He did infrared saunas five times per week to enhance detoxification, showering with nontoxic castile soap immediately thereafter. He bumped his fiber intake to 35 to 40 grams per day, which he followed on Cronometer, supplementing

Prescription for an Ageless Brain

with organic psyllium husk to get to his target of 35 to 40 grams. He included crucifers like brussels sprouts and cauliflower in his diet and took sulforaphane daily. He kept well hydrated with three to four liters of filtered water each day (including what he received from his fruits and vegetables).

He held back adding high-dose thiamine, methylene blue, and *Mucuna pruriens* because he did not yet have motor symptoms. However, he was concerned enough about his future that he talked with his physician about ADRCs, which are adipose-derived stem cells (autologous stem cells) that were harvested from him and then given to him, avoiding risk of infectious agents from other donors.

After discussion with his family and doctor, Carlton decided that his best hope for a long life free of symptoms of neurodegeneration would be to go ahead with the ADRCs. This turned out to be a relatively simple procedure, and from his own adipose tissue, just over 100 million stem cells were isolated and injected. Mannitol was given to open his blood-brain barrier transiently, allowing improved access of the cells to his brain. He had only a brief period of anesthesia, but nonetheless increased his glutathione, took vitamin C, and did sauna for the week following his anesthesia to ensure rapid detox from the anesthetic.

Over the next ten years, he continued his detoxification, continued optimizing his glutathione level, saunas each week, and overall healthy brain aging, with a plant-rich diet, daily exercise, optimizing his sleep, minimizing stress, doing some brain training, and continued his supplements targeting detox and mitochondrial energetics. His constipation disappeared, his sense of smell actually improved, and his RBD declined from several times per week to twice, then once every few weeks, and finally it only occurred about once per year, even skipping some years.

He reached seventy doing well and had a long talk with the physician who had helped him to avoid the dreadful march toward disability with Parkinson's. He had never developed the tremor, the tiny writing, the speech and swallowing problems, the falls, or any of the other major symptoms of Parkinson's. He had a repeat Syn-One biopsy, which showed only a minimal suggestion of positivity at one of the three sites—a clear improvement from the twenty-three years prior. His biological age was determined to be fifty-four.

This brush with the earliest symptoms of Parkinson's had taught him that he must avoid toxins, continue to support his genetically less-than-optimal detox system, and continue his healthy lifestyle. He also recognized that, although he had avoided the severe symptoms of Parkinson's, he was likely to be mildly or even moderately deficient in his major dopamine pathway—these Parkinson's symptoms do not occur until about 80 percent of the dopamine has been lost from the major motor modulation pathway, which is the nigrostriatal pathway (running from the substantia nigra in the brainstem to the striatum, which is deep in the brain's hemispheres), so he certainly could be 10 percent, 20 percent, or even 50 percent low and he must protect his brain. The RBD had been a life-changing wakeup call.

At seventy-seven, a screening carotid ultrasound showed that his left carotid had a 25 percent narrowing due to some mild atherosclerotic plaque. He started nattokinase and found the study showing that 10,000 units of nattokinase was associated with reduction in plaque, without any bleeding complications. Being cautious, he took only 6,000 units each day, but follow-ups over the next ten years showed no increase, smooth plaque, and no turbulent flow, so he was pleased with that.

At eighty-two, he had some difficulty with urination, which turned out to be due to prostatic hypertrophy. His PSA was normal. His doctor recommended a TURP (transurethral resection of the prostate) to improve his urine flow, and Carlton asked if this could be done via spinal anesthesia, to avoid general anesthesia. This was completed without incident, and the anesthesiologist finished by adding glutathione to his IV, to help with rapid detox from the spinal anesthetic. He had two episodes of RBD in the ensuing six months, but then these ceased once again.

At eighty-eight, he had some transient heart palpitations, which turned out to be due to a bout of atrial fibrillation. This was managed with the triad of taurine 1000 mg, electrolytes including potassium 500 mg, and magnesium 1000 mg each evening. With these on board, he had no further episodes. It turned out that this had been incited by some mild sleep apnea, which was corrected with a simple dental device.

Into his nineties, he especially enjoyed the walks along the water with his wife, recognizing that without his early treatment, he would not have been able to take such relaxing strolls. He told me how much he appreciated saving his brain, being able to live out his life sharp and unimpaired. He said, "Every moment that you can avoid a stumbling, confused brain is well worth your time for early detection and comprehensive brain care. Keeping a youthful brain has made all of the difference in my life, Dr. Bredesen. Could you get that message out to as many people as possible?"

That is an excellent point, Carlton, I'll do that.

ACKNOWLEDGMENTS

I am grateful to my wife, Aida, who is always focused on improving patients' lives—and to our beloved daughters, Tara and Tess. Special thanks to Diana Merriam and the Evanthea Foundation for their vision, commitment, continued enthusiasm, and guidance during two clinical trials. I am grateful to Phyllis and the late Jim Easton for their commitment to making a difference for people with Alzheimer's disease. I am also grateful to Katherine Gehl, Jessica Lewin, Wright Robinson, Dr. Patrick Soon-Shiong, Cary and Will Singleton, Douglas Rosenberg, Beryl Buck, Dagmar and David Dolby, Stephen D. Bechtel Jr., Lucinda Watson, Tom Marshall and the Joseph Drown Foundation, Bill Justice, Dave and Sheila Mitchell, Josh Berman, Ben and Shelly Chigier, Hideo Yamada, and Jeffrey Lipton.

I have been fortunate to train with leading scientists and physicians: Professors Stanley Prusiner, Mark Wrighton (Chancellor), Roger Sperry, Robert Collins, Robert Fishman, Roger Simon, Vishwanath Lingappa, William Schwartz, Kenneth McCarty Jr., J. Richard Baringer, Neil Raskin, Robert Layzer, Seymour Benzer, Erkki Ruoslahti, Lee Hood, and Mike Merzenich.

Special thanks to director Hideyuki Tokigawa, narrator Michael Bublé, and director of photography Ivan Kovac of the documentary *Memories for Life—Reversing Alzheimer's*.

I am also grateful to the functional medicine pioneers and experts who are revolutionizing medicine and healthcare: Drs. Jeffrey Bland, David Perlmutter, Mark Hyman, Dean Ornish, Ritchie Shoemaker, Neil Nathan, Joseph Pizzorno, Sara Gottfried, David Jones, Robert Lustig, Patrick Hanaway, Terry Wahls, Stephen Gundry, Ari Vojdani, Prudence Hall, Tom O'Bryan, Chris Kresser, Mary Kay Ross, Ann Hathaway, Kathleen Toups, Deborah Gordon, Jeralyn Brossfield, Kristine Burke, Jill Carnahan, Susan Sklar, Mary Ackerley, Sunjya Schweig, Sharon Hausman-Cohen, Nate Bergman, David Haase, Kim Clawson Rosenstein, Wes Youngberg, Craig Tanio, Hans Frykman, Jan Venter, Dave Jenkins, Miki Okuno, Elroy Vojdani, Chris Shade, health coaches Kerry and Timothy Rutland, Judy Benjamin, Robyn Albaum, Lisa Carson, Karen Malkin, Amylee Amos, Aarti Batavia, and Tess Bredesen, and the over two thousand physicians from ten countries and around the United States who have participated in and contributed to the course focused on the protocol described in this book. In addition, I am grateful to Lance Kelly, Julie Gregory, Sho Okada, Bill Lipa, Scott Grant, Ryan Morishige, Ekta Agrawal, Christine Coward, Jane Connelly, Lucy Kim, Melissa Manning, Casey Currie, Chase Kennedy, Gahren Markarian, and the team at Apollo Health for their outstanding work on the ReCODE algorithm, coding, and reports; to Craig Weston, Deb Geihsler, Howard Burde, Will Nields, and the team at Grey Matters; to Darrin Peterson and the team at LifeSeasons; to Taka Kondo and the team at Yamada Bee.

For three decades of experiments that led us to the first reversals of cognitive decline, I am grateful to Shahrooz Rabizadeh, Patrick Mehlen, Varghese John, Rammohan Rao, Patricia Spilman, Jesus Campagna, Rowena Abulencia, Kayvan Niazi, Litao Zhong, Alexei Kurakin, Darci Kane, Karen Poksay, Clare Peters-Libeu, Veena Theendakara, Veronica Galvan, Molly Susag, Alex Matalis, and all the

Acknowledgments

other members of the Bredesen Laboratory, as well as to my colleagues at the Buck Institute for Research on Aging, UCSF, the Sanford Burnham Prebys Medical Discovery Institute, and UCLA. I am also grateful to Dan Kelly, David Merrill, Neil Martin, and the superb team at the Pacific Neuroscience Institute for their vision to support the establishment of a program in precision brain health.

For their friendship and many discussions over the years, I thank Shahrooz Rabizadeh, Patrick Mehlen, Edwin and Chris Amos, Michael Ellerby, David Greenberg, John Reed, Guy Salvesen, Tuck Finch, Nuria Assa-Munt, Kim and Rob Rosenstein, Eric Tore and Carol Adolfson, Akane Yamaguchi, Judy and Paul Bernstein, Beverly and Roldan Boorman, Sandy and Harlan Kleiman, Philip Bredesen and Andrea Conte, Deborah Freeman, Peter Logan, Sandi and Bill Nicholson, Stephen and Mary Kay Ross, Mary McEachron, and Douglas Green.

Finally, I am grateful for the outstanding team with which I have worked on this book: for the writing and editing of Matthew LaPlante; literary agents John Maas and Celeste Fine of Park & Fine; the team at Amy Stanton & Co., and editor Julie Will of Flatiron Books.

NOTES

1: PERFORMANCE AND PROTECTION

1. Lawrence Growbel, "The Remarkable Dr. Feynman: Caltech's Eccentric Richard P. Feynman Is a Nobel Laureate, a Member of the Shuttle Commission, and Arguably the World's Best Theoretical Physicist," *Los Angeles Times*, April 20, 1986, https://www.latimes.com/archives/la-xpm-1986-04-20-tm-1265-story.html.
2. Kat Toups et al., "Precision Medicine Approach to Alzheimer's Disease: Successful Pilot Project," *Journal of Alzheimer's Disease* 88, no. 4 (2022): 1411–1421.
3. Kinga Igloi et al., "Interactions between Physical Exercise, Associative Memory, and Genetic Risk for Alzheimer's Disease," *Cerebral Cortex* 34, no. 5 (2024): bhae205.
4. L. Hood and N. Price, *The Age of Scientific Wellness: Why the Future of Medicine Is Personalized, Predictive, Data-Rich, and in Your Hands* (Harvard University Press, 2023).
5. A. Bonneville-Roussy et al., "Music through the Ages: Trends in Musical Engagement and Preferences from Adolescence through Middle Adulthood," *Journal of Personality and Social Psychology* 105, no. 4 (2013): 703.
6. Blue Cross Blue Shield, "Early-Onset Dementia and Alzheimer's Rates Grow for Younger Americans," February 27, 2020, https://www.bcbs.com/dA%20/bb22aac725/fileAsset/HOA-Early-Onset-Dementia-Alzheimers_2020.pdf.
7. W. R. Powell et al., "Association of Neighborhood-Level Disadvantage with Alzheimer Disease Neuropathology," *JAMA Network Open* 3, no. 6 (2020): e207559.

8. S. Hendriks et al., "Global Prevalence of Young-Onset Dementia: A Systematic Review and Meta-Analysis," *JAMA Neurology* 78, no. 9 (2021): 1080–1090.
9. C. Delaby et al., "Overview of the Blood Biomarkers in Alzheimer's Disease: Promises and Challenges," *Revue Neurologique* 179, no. 3 (2023): 161–172.
10. L. A. Manwell et al., "Digital Dementia in the Internet Generation: Excessive Screen Time during Brain Development Will Increase the Risk of Alzheimer's Disease and Related Dementias in Adulthood," *Journal of Integrative Neuroscience* 21, no. 1 (2022): 28.
11. C. L. Tsai et al., "Differences in Neurocognitive Performance and Metabolic and Inflammatory Indices in Male Adults with Obesity as a Function of Regular Exercise," *Experimental Physiology* 104, no. 11 (2019): 1650–1660.
12. Richard Dawkins, *The Selfish Gene* (Oxford University Press, 2016).
13. George C. Williams, "Pleiotropy, Natural Selection, and the Evolution of Senescence," *Evolution* 11 (1957): 398–411, republished in *Science of Aging Knowledge Environment* 2001, no. 1 (2001): cp13.
14. Josh Mitteldorf, "What Is Antagonistic Pleiotropy?," *Biochemistry* (Moscow) 84, no. 12 (2019): 1458–1468.
15. Richard Feynman, *The Character of Physical Law* (BBC, 1965; repr., Cox and Wyman Ltd., 1967).
16. Bastiaan R. Bloem and Tjitske A. Boonstra, "The Inadequacy of Current Pesticide Regulations for Protecting Brain Health: The Case of Glyphosate and Parkinson's Disease." *Lancet Planetary Health* 7, no. 12 (2023): e948–e949.
17. Burak Yulug et al., "Combined Metabolic Activators Improve Cognitive Functions in Alzheimer's Disease Patients: A Randomised, Double-Blinded, Placebo-Controlled Phase-II Trial," *Translational Neurodegeneration* 12, no. 1 (2023): 4.
18. Jing Yuan et al., "Is Dietary Choline Intake Related to Dementia and Alzheimer's Disease Risks? Results from the Framingham Heart Study," *American Journal of Clinical Nutrition* 116, no. 5 (2022): 1201–1207.
19. Dennis J. Selkoe and John Hardy, "The Amyloid Hypothesis of Alzheimer's Disease at 25 Years," *EMBO Molecular Medicine* 8, no. 6 (2016): 595–608, https://doi.org/10.15252/emmm.201606210.

2: ADDING INSULTS TO AGING

1. Raymond D. Palmer, "Three Tiers to Biological Escape Velocity: The Quest to Outwit Aging," *Aging Medicine* 5, no. 4 (2022): 281–286, https://doi.org/10.1002/agm2.12231.
2. M. Ackermann et al., "On the Evolutionary Origin of Aging," *Aging Cell* 6, no. 2 (2007): 235–244.

3. F. William Danby, "Nutrition and Aging Skin: Sugar and Glycation," *Clinics in Dermatology* 28, no. 4 (2010): 409–411.
4. Pouya Saeedi et al., "Global and Regional Diabetes Prevalence Estimates for 2019 and Projections for 2030 and 2045: Results from the International Diabetes Federation Diabetes Atlas, 9th edition," *Diabetes Research and Clinical Practice* 157 (2019): 107843, https://doi.org/10.1016j.diabres.2019.107843.
5. Gary Taubes, "Is Sugar Toxic?," *New York Times Magazine*, April 13, 2011.
6. D. Kellar and S. Craft, "Brain Insulin Resistance in Alzheimer's Disease and Related Disorders: Mechanisms and Therapeutic Approaches," *Lancet Neurology* 19, no. 9 (2020): 758–766.
7. C. B. Amidei et al., "Association between Age at Diabetes Onset and Subsequent Risk of Dementia," *JAMA* 325, no. 16 (2021): 1640–1649.
8. Richard J. Johnson et al., "Could Alzheimer's Disease Be a Maladaptation of an Evolutionary Survival Pathway Mediated by Intracerebral Fructose and Uric Acid Metabolism?," *American Journal of Clinical Nutrition* 117, no. 3 (2023): 455–466, https://doi.org/10.1016/j.ajcnut.2023.01.00.
9. Alejandro Gugliucci, "Formation of Fructose-Mediated Advanced Glycation End Products and their Roles in Metabolic and Inflammatory Diseases," *Advances in Nutrition* 8, no. 1 (2017): 54–62, https://doi.org/10.3945/an.116.013912.
10. B. Manivannan et al., "Assessment of Persistent, Bioaccumulative and Toxic Organic Environmental Pollutants in Liver and Adipose Tissue of Alzheimer's Disease Patients and Age-Matched Controls," *Current Alzheimer Research* 16, no. 11 (2019): 1039–1049.
11. Jae-Hyun Yang et al., "Loss of Epigenetic Information As a Cause of Mammalian Aging," *Cell* 186, no. 2 (2023): 305–326.
12. M. Manikkam et al., "Transgenerational Actions of Environmental Compounds on Reproductive Disease and Identification of Epigenetic Biomarkers of Ancestral Exposures," *PLoS One* 7, no. 2 (2012): e31901.
13. S. C. Burgess and D. J. Marshall, "Adaptive Parental Effects: The Importance of Estimating Environmental Predictability and Offspring Fitness Appropriately," *Oikos* 123, no. 7 (2014): 769–776.
14. S. Jiang et al., "Epigenetic Modifications in Stress Response Genes Associated with Childhood Trauma," *Frontiers in Psychiatry* 10 (2019): 808.
15. Germán Alberto Nolasco-Rosales et al., "Aftereffects in Epigenetic Age Related to Cognitive Decline and Inflammatory Markers in Healthcare Personnel with Post-COVID-19: A Cross-Sectional Study," *International Journal of General Medicine* 16 (2023): 4953–4964.
16. Natalie C. Silmon de Monerri and Kami Kim, "Pathogens Hijack the

Epigenome: A New Twist on Host-Pathogen Interactions," *American Journal of Pathology* 184, no. 4 (2014): 897–911.

17. S. Dubey et al., "The Effects of SARS-CoV-2 Infection on the Cognitive Functioning of Patients with Pre-Existing Dementia," *Journal of Alzheimer's Disease Reports* 7, no. 1 (2023): 119–128.

18. A. H. Bayat et al., "COVID-19 Causes Neuronal Degeneration and Reduces Neurogenesis in Human Hippocampus," *Apoptosis* 27, nos. 11–12 (2022): 852–868.

19. Thiruselvam Viswanathan et al., "Structural Basis of RNA Cap Modification by SARS-CoV-2," *Nature Communications* 11, no. 1 (2020): 3718, https://doi.org/10.1038/s41467-020-17496-8.

20. Leah S. Richmond-Rakerd et al., "Associations of Hospital-Treated Infections with Subsequent Dementia: Nationwide 30-Year Analysis," *Nature Aging* 4, no. 6 (2024): 1–8, https://doi.org/10.1038/s43587-024-00621-3.

21. P. S. Stein et al., "Serum Antibodies to Periodontal Pathogens Are a Risk Factor for Alzheimer's Disease," *Alzheimer's & Dementia* 8, no. 3 (2012): 196–203.

22. R. Sender, S. Fuchs, and R. Milo, "Revised Estimates for the Number of Human and Bacteria Cells in the Body," *PLoS Biology* 14, no. 8 (2016): e1002533.

23. R. Khan, F. C. Petersen, and S. Shekhar, "Commensal Bacteria: An Emerging Player in Defense against Respiratory Pathogens," *Frontiers in Immunology* 10 (2019): 1203, https://doi.org/10.3389/fimmu.2019.01203.

24. J. A. Gilbert and J. D. Neufeld, "Life in a World without Microbes," *PLoS Biology* 12, no. 12 (2014): https://doi.org/10.1371/journal.pbio.1002020.

25. Andrei B. Borisov, Shi-Kai Huang, and Bruce M. Carlson, "Remodeling of the Vascular Bed and Progressive Loss of Capillaries in Denervated Skeletal Muscle," *Anatomical Record* 258, no. 3 (2000): 292–304, https://doi.org/10.1002/(SICI)1097-0185(20000301)258:3<292::AID-AR9>3.0.CO;2-N.

26. Sabrina S. Salvatore, Kyle N. Zelenski, and Ryan K. Perkins, "Age-Related Changes in Skeletal Muscle Oxygen Utilization," *Journal of Functional Morphology and Kinesiology* 7, no. 4 (2022): 87, https://doi.org/10.3390/jfmk7040087.

27. Dimitry A. Chistiakov et al., "Mitochondrial Aging and Age-Related Dysfunction of Mitochondria," special issue, *BioMed Research International* (2014), https://doi.org/10.1155/2014/238463.

28. Paolo Tessari, "Changes in Protein, Carbohydrate, and Fat Metabolism with Aging: Possible Role of Insulin," *Nutrition Reviews* 58, no. 1 (2000): 11–19, https://doi.org/10.1111/j.1753-4887.2000.tb01819.x.

29. A. Miller, "Clear Lake Man Finds 'Freedom' from ALS from Cycling," *Des Moines Register*, July 4, 2014. *Note*: Humberg died on September 20, 2021, at his home in Clear Lake, Iowa, surrounded by family and friends, more than fifteen years after his ALS diagnosis.

Notes

30. W. Barrie et al., "Elevated Genetic Risk for Multiple Sclerosis Emerged in Steppe Pastoralist Populations," *Nature* 625, no. 7994 (2024): 321–328, https://doi.org/10.1038/s41586-023-06618-z.
31. Tobias V. Lanz et al., "Clonally Expanded B Cells in Multiple Sclerosis Bind EBV EBNA1 and GlialCAM," *Nature* 603, no. 7900 (2022): 321–327, https://doi.org/10.1038/s41586-022-04432-7.
32. Andreas Yiallouris et al., "Adrenal Aging and Its Implications on Stress Responsiveness in Humans," *Frontiers in Endocrinology* 10 (2019): 54, https://doi.org/10.3389/fendo.2019.00054.

3: WHAT IS POSSIBLE AT ONE HUNDRED AND BEYOND?

1. Juan Fortea et al., "APOE4 Homozygozity Represents a Distinct Genetic Form of Alzheimer's Disease," *Nature Medicine* 30 (2024): 1284–1291, https://doi.org/10.1038/s41591-024-02931-w.
2. David A. Sinclair and Matthew D. LaPlante, *Lifespan: Why We Age—And Why We Don't Have To* (Atria Books, 2019).
3. Peter Diamandis, "Living to 200 Years Old: Unlocking the Secrets of the Bowhead Whale," Diamandis.com, June 11, 2023, https://www.diamandis.com/blog/bowhead-whale.
4. Michael Keane et al., "Insights into the Evolution of Longevity from the Bowhead Whale Genome," *Cell Reports* 10, no. 1 (2015): 112–122, https://linkinghub.elsevier.com/retrieve/pii/S2211-1247(14)01019-5.
5. Julius Nielsen et al., "Eye Lens Radiocarbon Reveals Centuries of Longevity in the Greenland Shark (*Somniosus microcephalus*)," *Science* 353, no. 6300 (2016): 702–704, https://www.science.org/doi/abs/10.1126/science.aaf1703.
6. Barry E. Flanary and Gunther Kletetschka, "Analysis of Telomere Length and Telomerase Activity in Tree Species of Various Life-Spans, and with Age in the Bristlecone Pine Pinus longaeva," *Biogerontology* 6 (2005): 101–111, https://doi.org/10.1007/s10522-005-3484-4.
7. A. A. Lisenkova et al., "Complete Mitochondrial Genome and Evolutionary Analysis of *Turritopsis dohrnii*, the 'Immortal' Jellyfish with a Reversible Life-Cycle," *Molecular Phylogenetics and Evolution* 107 (2017): 232–238, https://doi.org/10.1016/j.ympev.2016.11.007.
8. Linda Dieckmann et al., "Characteristics of Epigenetic Aging across Gestational and Perinatal Tissues," *Clinical Epigenetics* 13, no. 1 (2021): 1–17, https://doi.org/10.1186/s13148-021-01080-y.
9. D. William Molloy and Timothy I. M. Standish, "A Guide to the Standardized Mini-Mental State Examination," *International Psychogeriatrics* 9, no. S1 (1997): 87–94, https://doi.org/10.1017/S1041610297004754.
10. N. Beker et al., "Neuropsychological Test Performance of Cognitively Healthy

Centenarians: Normative Data from the Dutch 100-Plus Study," *Journal of the American Geriatrics Society* 67, no. 4 (2019): 759–767.

11. Harrison Jones, "Hanover's Iron Man: Les Savino Still Going Strong at 100," *Evening Sun* (Hanover, PA), August 30, 2022, https://www.eveningsun.com/story/news/local/2022/08/30/100th-birthday-hanover-pa-les-savino-celebrated-longevity-hanover-ymca/65461530007/.

12. Kayleigh Johnson, "100-Year-Old Man Refuses to 'Give Up,' Works Out at YMCA Every Day," FOX43, September 7, 2022, https://www.fox43.com/video/news/local/york-county/100-year-old-man-works-out-hanover-ymca-les-savino/521-955e311a-7384-4d8-9fc9-0ea877d21990.

13. Gretchen Cuda Kroen, "At 100 and Recognized as the World's Oldest Practicing Physician, This Cleveland Doctor Is Still Going Strong," Cleveland.com, September 17, 2022, https://www.cleveland.com/news/2022/09/at-100-and-recognized-as-the-worlds-oldest-practicing-physician-this-cleveland-doctor-is-still-going-strong.html.

14. Ester Bloom, "100-Year-Old Sisters Share 4 Tips for Staying Mentally Sharp As You Age—and They Don't Say Crossword Puzzles," CNBC, March 27, 2023, https://www.cnbc.com/2023/03/27/100-year-old-sisters-share-tips-for-staying-mentally-sharp-as-you-age.html.

15. Katherine Schaeffer, "U.S. Centenarian Population Is Projected to Quadruple over the Next 30 Years," Pew Research Center, January 9, 2024, https://www.pewresearch.org/short-reads/2024/01/09/us-centenarian-population-is-projected-to-quadruple-over-the-next-30-years/#:~:text=Centenarians%20around%20the%20world,than%20the%20Census%20Bureau's%20estimate\.

16. Tiia Ngandu et al., "A 2 Year Multidomain Intervention of Diet, Exercise, Cognitive Training, and Vascular Risk Monitoring versus Control to Prevent Cognitive Decline in At-Risk Elderly People (FINGER): A Randomised Controlled Trial," *Lancet* 385, no. 9984 (2015): 2255–2263, https://doi.org/10.1016/S0140-6736(15)60461-5.

17. Andrew Sommerlad et al., "Social Participation and Risk of Developing Dementia," *Nature Aging* 3, no. 5 (2023): 532–545, https://doi.org/10.1038/s43587-023-00387-0.

18. Oliver M. Shannon et al., "Mediterranean Diet Adherence Is Associated with Lower Dementia Risk, Independent of Genetic Predisposition: Findings from the UK Biobank Prospective Cohort Study," *BMC Medicine* 21, no. 1 (2023): 1–13, https://doi.org/10.1186/s12916-023-02772-3.

19. May A. Beydoun, H. A. Beydoun, and Youfa Wang, "Obesity and Central Obesity As Risk Factors for Incident Dementia and Its Subtypes: A System-

atic Review and Meta-Analysis," *Obesity Reviews* 9, no. 3 (2008): 204–218, https://doi.org/10.1111/j.1467-789X.2008.00473.x.

20. Soo Borson et al., "Improving Dementia Care: The Role of Screening and Detection of Cognitive Impairment," *Alzheimer's & Dementia* 9, no. 2 (2013): 151–159, https://doi.org/10.1016/j.jalz.2012.08.008.

21. Joseph E. Ebinger et al., "Association of Blood Pressure Variability during Acute Care Hospitalization and Incident Dementia," *Frontiers in Neurology* 14 (2023): 1085885, https://doi.org/10.3389/fneur.2023.1085885.

22. David A. Sbarra, Rita W. Law, and Robert M. Portley, "Divorce and Death: A Meta-Analysis and Research Agenda for Clinical, Social, and Health Psychology," *Perspectives on Psychological Science* 6, no. 5 (2011): 454–474, https://doi.org/10.1177/1745691611414724.

23. M. G. Griswold et al., "Alcohol Use and Burden for 195 Countries and Territories, 1990–2016: A Systematic Analysis for the Global Burden of Disease Study 2016," *Lancet*, 392, no. 10152 (2018): 1015–1035, https://doi.org/10.1016/S0140-6736(18)31310-2.

24. Carl Haub, "How Many People Have Ever Lived on Earth?," *Population Today* 23, no. 2 (1995): 4–5, https://www.safetylit.org/citations/index.php?fuseaction=citations.viewdetails&citationIds[]=citjournalarticle_209327_38.

25. Anthony Medford and James W. Vaupel, "Human Lifespan Records Are Not Remarkable but Their Durations Are," *PloS One* 14, no. 3 (2019): e0212345, https://doi.org/10.1371/journal.pone.0212345.

26. Vyara Todorova and Arjan Blokland, "Mitochondria and Synaptic Plasticity in the Mature and Aging Nervous System," *Current Neuropharmacology* 15, no. 1 (2017): 166–173, https://doi.org/10.2174/1570159x14666160414111821.

27. Soyon Hong et al., "Complement and Microglia Mediate Early Synapse Loss in Alzheimer Mouse Models," *Science* 352, no. 6286 (2016): 712–716, https://doi.org/10.1126/science.aad8373.

28. Donna Vickroy, "'The Things You Remember': Centenarians Share What It's Like to Be 100," *Chicago Tribune*, October 23, 2017, https://www.chicagotribune.com/2017/10/23/the-things-you-remember-centenarians-share-what-its-like-to-be-100/.

29. Dylan Loeb McClain, "Yuri Averbakh, Chess's First Centenarian Grandmaster, Dies at 100," *New York Times*, May 9, 2022, https://www.nytimes.com/2022/05/09/sports/yuri-averbakh-dead.html.

30. Peter Doggers, "Yuri Averbakh, the Oldest Living Grandmaster, Turns 100," *Chess*, February 8, 2022, https://www.chess.com/news/view/yuri-averbakh-100-years.

31. "100 Years Young and Still Going Strong . . . Thanks to her Nintendo!

Pensioner Clocks Up a Century and Put Her Quick Wits Down to Handheld Console," *Daily Mail*, February 1, 2012, https://www.dailymail.co.uk/health/article-2094402/Kathleen-Connell-100-puts-quick-wits-Nintendo-DS-Lite.html.

32. Philip Townsend, "Norfolk's Stanley Sacks Is the Oldest Practicing Attorney in the Country," 13newsnow, January 20, 2023, https://www.13newsnow.com/article/features/norfolks-stanley-sacks-oldest-practicing-attorney/291-146cacd2-3497-4060-89c2-66b95dfbb384.

33. Paul Laity, "The 100-Year-Old Couple—Still Married, Still Going Strong," *Guardian*, February 11, 2017, https://www.theguardian.com/lifeandstyle/2017/feb/11/the-100-year-old-couple-still-married-still-going-strong.

34. Jasmin Aline Persch, "At Age 102, This Therapist Is Still Psyched," *Today*, November 14, 2011, https://www.today.com/news/age-102-therapist-still-psyched-wbna45293812.

35. A. Pawlowski, "At 100, She Loves to Dance the Tush Push, Do Yoga, Eat Chocolate Every Day," *Today*, https://www.today.com/health/womens-health/longevity-advice-dancing-nana-100-years-old-rcna77660.

36. Simona Lattanzi et al., "Blood Pressure Variability Predicts Cognitive Decline in Alzheimer's Disease Patients," *Neurobiology of Aging* 35, no. 10 (2014): 2282–2287, https://doi.org/10.1016/j.neurobiolaging.2014.04.023.

4: DYING OF PROFIT

1. Cameron Langford, "Fifth Circuit Rejects Lyme Disease Patients' Coverage Denial Conspiracy Claims," Courthouse News Service, November 16, 2023, https://www.courthousenews.com/fifth-circuit-rejects-lyme-disease-patients-coverage-denial-conspiracy-claims.

2. Kaare Christensen et al., "Ageing Populations: The Challenges Ahead," *Lancet* 374, no. 9696 (2009): 1196–1208, https://doi.org/10.1016/S0140-6736(09)61460-4.

3. Vasilis Kontis et al., "Future Life Expectancy in 35 Industrialised Countries: Projections with a Bayesian Model Ensemble," *Lancet* 389, no. 10076 (2017): 1323–1335, https://doi.org/10.1016/S0140-6736(16)32381-9.

4. Dale Bredesen, *The First Survivors of Alzheimer's: How Patients Recovered Life and Hope in Their Own Words* (Avery, 2021).

5. Markku Kurkinen, "Lecanemab (Leqembi) Is Not the Right Drug for Patients with Alzheimer's Disease," *Advances in Clinical and Experimental Medicine* 32, no. 9 (2023): 943–947, https://doi.org/10.17219/acem/171379.

6. Dale E. Bredesen et al., "Reversal of Cognitive Decline in Alzheimer's Disease," *Aging* (Albany, NY) 8, no. 6 (2016): 1250, https://doi.org/10.18632/aging.100981.

7. Melody Petersen, "Inside the Plan to Diagnose Alzheimer's in People with No Memory Problems and Who Stands to Benefit," *Los Angeles Times*, February 19, 2024, https://www.latimes.com/science/story/2024-02-14/inside-controversial-plan-to-diagnose-alzheimers-in-people-without-symptoms.
8. Tom Nicholson, "Where Is Painkiller's Richard Sackler Now?," *Esquire*, August 10, 2023.
9. Evan Hughes, "The Pain Hustlers," *New York Times Magazine*, May 2, 2018, https://www.nytimes.com/interactive/2018/05/02/magazine/money-issue-insys-opioids-kickbacks.html.
10. Theodore J. Cicero et al., "The Changing Face of Heroin Use in the United States: A Retrospective Analysis of the Past 50 Years," *JAMA Psychiatry* 71, no. 7 (2014): 821–826, https://doi.org/10.1001/jamapsychiatry.2014.366.
11. Catherine Shoard, "Back to the Future Day: What Part II Got Right and Wrong about 2015—an A–Z," *Guardian*, October 20, 2015, https://www.theguardian.com/film/filmblog/2015/jan/02/what-back-to-the-future-part-ii-got-right-and-wrong-about-2015-an-a-z.
12. Mark Erickson, "The Science of Doctor Who," in *Doctor Who and Science: Essays on Ideas, Identities and Ideologies in the Series*, ed. Marcus K. Harmes and Lindy A. Orthia (TK McFarland, 2021), 205.

5: IDENTIFY YOUR WHY

1. Caroline A. Koch, Emilie W. Kjeldsen, and Ruth Frikke-Schmidt, "Vegetarian or Vegan Diets and Blood Lipids: A Meta-Analysis of Randomized Trials," *European Heart Journal* 44, no. 28 (2023): 2609–26122, https://doi.org/10.1093/eurheartj/ehad211.
2. Byron J. Hoogwerf, "Statins May Increase Diabetes, but Benefit Still Outweighs Risk," *Cleveland Clinic Journal of Medicine* 90, no. 1 (2023): 53–62.
3. Evelyn Medawar et al., "The Effects of Plant-Based Diets on the Body and the Brain: A Systematic Review," *Translational Psychiatry* 9, no. 1 (2019): 226.
4. Seung Hee Lee et al., "Adults Meeting Fruit and Vegetable Intake Recommendations—United States, 2019," *Morbidity and Mortality Weekly Report* 71, no. 1 (2022): 1, https://www.cdc.gov/mmwr/volumes/71/wr/mm7101a1.htm.
5. Alana Rhone et al., "Low-Income and Low-Supermarket-Access Census Tracts, 2010–2015," *Economic Information Bulletin* 165 (USDA, Economic Research Service, January 2017).
6. Andrea Carlson and Elizabeth Frazão, "Are Healthy Foods Really More Expensive? It Depends on How You Measure the Price," *Economic Information Bulletin* 96 (USDA, Economic Research Service, May 2012).

7. Martin Loef and Harald Walach, "Midlife Obesity and Dementia: Meta-Analysis and Adjusted Forecast of Dementia Prevalence in the United States and China," *Obesity* 21, no. 1 (2013): E51–E55.
8. Maria Ly et al., "Neuroinflammation: A Modifiable Pathway Linking Obesity, Alzheimer's Disease, and Depression," *American Journal of Geriatric Psychiatry* (2023): 853–856, https://doi.org/10.1016/j.jagp.2023.06.001.
9. Milad Kheirvari et al., "The Changes in Cognitive Function Following Bariatric Surgery Considering the Function of Gut Microbiome," *Obesity Pillars* 3 (2022): 100020.
10. Moein Askarpour et al., "Effect of Bariatric Surgery on Serum Inflammatory Factors of Obese Patients: A Systematic Review and Meta-Analysis," *Obesity Surgery* 29 (2019): 2631–2647, https://doi.org/10.1007/s11695-019-03926-0.
11. Mohammed S. Ellulu et al., "Obesity and Inflammation: The Linking Mechanism and the Complications," *Archives of Medical Science* 13, no. 4 (2017): 851–863.
12. Jefferson W. Kinney et al., "Inflammation as a Central Mechanism in Alzheimer's Disease," *Alzheimer's & Dementia* 4, no. 1 (2018): 575–590.
13. Andrew Steptoe, Mark Hamer, and Yoichi Chida, "The Effects of Acute Psychological Stress on Circulating Inflammatory Factors in Humans: A Review and Meta-Analysis," *Brain, Behavior, and Immunity* 21, no. 7 (2007): 901–912.
14. Michael R. Irwin, Richard Olmstead, and Judith E. Carroll, "Sleep Disturbance, Sleep Duration, and Inflammation: A Systematic Review and Meta-Analysis of Cohort Studies and Experimental Sleep Deprivation," *Biological Psychiatry* 80, no. 1 (2016): 40–52.
15. Armin Imhof et al., "Effect of Alcohol Consumption on Systemic Markers of Inflammation," *Lancet* 357, no. 9258 (2001): 763–767.
16. Priyanka Chatterjee et al., "Evaluation of Anti-Inflammatory Effects of Green Tea and Black Tea: A Comparative *in vitro* Study," *Journal of Advanced Pharmaceutical Technology & Research* 3, no. 2 (2012): 136.
17. Charles N. Serhan, "Pro-Resolving Lipid Mediators Are Leads for Resolution Physiology," *Nature* 510, no. 7503 (2014): 92–101, https://doi.org/10.1038/nature13479.
18. Sonya R. Hardin, "Cat's Claw: An Amazonian Vine Decreases Inflammation in Osteoarthritis," *Complementary Therapies in Clinical Practice* 13, no. 1 (2007): 25–28, https://doi.org/10.1016/j.ctcp.2006.10.003.
19. Subathra Murugan et al., "The Neurosteroid Pregnenolone Promotes Degradation of Key Proteins in the Innate Immune Signaling to Suppress Inflammation," *Journal of Biological Chemistry* 294, no. 12 (2019): 4596–4607, https://doi.org/10.1074/jbc.RA118.005543.

Notes

20. Jiao Wang et al., "Poor Pulmonary Function Is Associated with Mild Cognitive Impairment, Its Progression to Dementia, and Brain Pathologies: A Community-Based Cohort Study," *Alzheimer's & Dementia* 18, no. 12 (2022): 2551–2559.
21. Qing Meng, Muh-Shi Lin, and I-Shiang Tzeng, "Relationship between Exercise and Alzheimer's Disease: A Narrative Literature Review," *Frontiers in Neuroscience* 14 (2020): 131.
22. Camilla Pellegrini et al., "A Meta-Analysis of Brain DNA Methylation across Sex, Age, and Alzheimer's Disease Points for Accelerated Epigenetic Aging in Neurodegeneration," *Frontiers in Aging Neuroscience* 13 (2021): 639428.
23. Gregory M. Fahy et al., "Reversal of Epigenetic Aging and Immunosenescent Trends in Humans," *Aging Cell* 18, no. 6 (2019): e13028, https://doi.org/10.1111/acel.13028.
24. Kara N. Fitzgerald et al., "Potential Reversal of Epigenetic Age Using a Diet and Lifestyle Intervention: A Pilot Randomized Clinical Trial," *Aging* (Albany, NY) 13, no. 7 (2021): 9419, https://doi.org/10.18632/aging.202913.
25. Borut Poljšak et al., "The Central Role of the NAD+ Molecule in the Development of Aging and the Prevention of Chronic Age-Related Diseases: Strategies for NAD+ Modulation," *International Journal of Molecular Sciences* 24, no. 3 (2023): 2959.
26. Oprah Winfrey, "Every Person Has a Purpose," *O, The Oprah Magazine*, October 2009, https://www.oprah.com/spirit/how-oprah-winfrey-found-her-purpose.
27. Hanns Hippius and Gabriele Neundörfer, "The Discovery of Alzheimer's Disease," *Dialogues in Clinical Neuroscience* 5, no. 1 (2003): 101–108, https://doi.org/10.31887/DCNS.2003.5.1/hhippius.
28. Filippo Cieri and Roberto Esposito, "Psychoanalysis and Neuroscience: The Bridge between Mind and Brain," *Frontiers in Psychology* 10 (2019): 465260, https://doi.org/10.3389/fpsyg.2019.01983.
29. Nicholas G. Norwitz et al., "Ketogenic Diet as a Metabolic Treatment for Mental Illness," *Current Opinion in Endocrinology, Diabetes and Obesity* 27, no. 5 (2020): 269–274, https://doi.org/10.1097/MED.0000000000000564.
30. Oleg Yerstein et al., "Benson's Disease or Posterior Cortical Atrophy, Revisited," *Journal of Alzheimer's Disease* 82, no. 2 (2021): 493–502.
31. "Posterior Cortical Atrophy," Mayo Clinic, https://www.mayoclinic.org/diseases-conditions/posterior-cortical-atrophy/diagnosis-treatment/drc-20376563.
32. Dale E. Bredesen, "Inhalational Alzheimer's Disease: An Unrecognized—and Treatable—Epidemic," *Aging* (Albany, NY) 8, no. 2 (2016): 304, https://doi.org/10.18632/aging.100896.

33. Sarah E. Jackson, Andrew Steptoe, and Jane Wardle, "The Influence of Partner's Behavior on Health Behavior Change: The English Longitudinal Study of Ageing," *JAMA Internal Medicine* 175, no. 3 (2015): 385–392.

6: A MEASURED APPROACH

1. Shridhara Alva et al., "Feasibility of Continuous Ketone Monitoring in Subcutaneous Tissue Using a Ketone Sensor," *Journal of Diabetes Science and Technology* 15, no. 4 (221): 768–774, https://www.ncbi.nlm.nih.gov/pmc/articles/PMC8252149/.
2. Stephen Cunnane, "Brain Energy Rescue with Ketones Improves Cognitive Outcomes in MCI," *Alzheimer's & Dementia* 18 (2022): e059627, https://doi.org/10.1002/alz.059627.
3. Bing Zhu et al., "HbA1c as a Screening Tool for Ketosis in Patients with Type 2 Diabetes Mellitus," *Scientific Reports* 6, no. 1 (2016): 39687, https://doi.org/10.1038/srep39687.
4. Chochanon Moonla et al., "Continuous Ketone Monitoring Via Wearable Microneedle Patch Platform," *ACS Sensors* 9, no. 2 (2024): 1004–1013, https://doi.org/10.1021/acssensors.3c02677.
5. Johannes Attems and Kurt A. Jellinger, "The Overlap between Vascular Disease and Alzheimer's Disease-Lessons from Pathology," *BMC Medicine* 12 (2014): 1–12, https://doi.org/10.1186/s12916-014-0206-2.
6. D. S. Knopman, "Cerebrovascular Disease and Dementia," special issue, *British Journal of Radiology* 80, no. 2 (2007): S121–S127, https://doi.org/10.1259/bjr/75681080.
7. Wiesje M. van der Flier et al., "Vascular Cognitive Impairment," *Nature Reviews Disease Primers* 4, no. 1 (2018): 1–16, https://doi.org/10.1038/nrdp.2018.3.
8. Jennifer Behbodikhah et al., "Apolipoprotein B and Cardiovascular Disease: Biomarker and Potential Therapeutic Target," *Metabolites* 11, no. 10 (2021): 690, https://doi.org/10.3390/metabo11100690.
9. Sara Kaffashian et al., "Predictive Utility of the Framingham General Cardiovascular Disease Risk Profile for Cognitive Function: Evidence from the Whitehall II Study," *European Heart Journal* 32, no. 18 (2011): 2326–2332, https://doi.org/10.1093/eurheartj/ehr133.
10. Amanda M. Perak et al., "Trends in Levels of Lipids and Apolipoprotein B in US Youths Aged 6 to 19 Years, 1999–2016," *JAMA* 321, no. 19 (2019): 1895–1905, https://doi.org/10.1001/jama.2019.4984.
11. Ian Galea, "The Blood–Brain Barrier in Systemic Infection and Inflammation," *Cellular & Molecular Immunology* 18, no. 11 (2021): 2489–2501, https://doi.org/10.1038/s41423-021-00757-x.

12. Lilly Shanahan, Jason Freeman, and Shawn Bauldry, "Is Very High C-Reactive Protein in Young Adults Associated with Indicators of Chronic Disease Risk?," *Psychoneuroendocrinology* 40 (2014): 76–85, https://doi.org/10.1016/j.psyneuen.2013.10.019.
13. Simona Luzzi et al., "Homocysteine, Cognitive Functions, and Degenerative Dementias: State of the Art," *Biomedicines* 10, no. 11 (2022): 2741, https://doi.org/10.3390/biomedicines10112741.
14. Ahmed Abdelhak et al., "Blood GFAP As an Emerging Biomarker in Brain and Spinal Cord Disorders," *Nature Reviews Neurology* 18, no. 3 (2022): 158–172.
15. Charlotte Johansson et al., "Plasma Biomarker Profiles in Autosomal Dominant Alzheimer's Disease," *Brain* 146, no. 3 (2023): 1132–1140, https://doi.org/10.1093/brain/awac399.
16. Tingting Liu et al., "Cerebrospinal Fluid GFAP Is a Predictive Biomarker for Conversion to Dementia and Alzheimer's Disease-Associated Biomarkers Alterations among de novo Parkinson's Disease Patients: A Prospective Cohort Study," *Journal of Neuroinflammation* 20, no. 1 (2023): 167.
17. Nicholas J. Ashton et al., "Diagnostic Accuracy of a Plasma Phosphorylated Tau 217 Immunoassay for Alzheimer Disease Pathology," *JAMA Neurology* 81, no. 3 (2024): 255–263, https://doi.org/10.1001/jamaneurol.2023.5319.
18. Matt Vasilogambros, "The NFL's Concussion Cover-Up," *Atlantic*, May 23, 2016.
19. Leah H. Rubin et al., "NFL Blood Levels Are Moderated by Subconcussive Impacts in a Cohort of College Football Players," *Brain Injury* 33, no. 4 (2019): 456–462.
20. Lenise A. Cummings-Vaughn et al., "Veterans Affairs Saint Louis University Mental Status Examination Compared with the Montreal Cognitive Assessment and the Short Test of Mental Status," *Journal of the American Geriatrics Society* 62, no. 7 (2014): 1341–1346, https://doi.org/10.1111/jgs.12874.

7: EATING FOR AN AGELESS BRAIN

1. S. E. Jacobsen and S. Sherwood, *Cultivo de granos andinos en Ecuador: Informe sobre los rubros quinua, chocho y amaranto* (Quito, Ecuador: Centro Internacional de la Papa, 2002), cited in Catherine Greene, "Organic Market Overview" (USDA, Economic Research Service, 2012).
2. Lindsay J. Collin et al., "Association of Sugary Beverage Consumption with Mortality Risk in US Adults: A Secondary Analysis of Data from the REGARDS Study," *JAMA Network Open* 2, no. 5 (2019): e193121, https://doi.org/10.1001/jamanetworkopen.2019.3121.
3. Carrie H. S. Ruxton and Madeleine Myers, "Fruit Juices: Are They Helpful

or Harmful? An Evidence Review," *Nutrients* 13, no. 6 (2021): 1815, https://doi.org/10.3390/nu13061815.

4. Belinda S. Lennerz et al., "Behavioral Characteristics and Self-Reported Health Status among 2029 Adults Consuming a 'Carnivore Diet,'" *Current Developments in Nutrition* 5, no. 12 (2021): nzab133, https://doi.org/10.1093/cdn/nzab133.

5. Serge H. Ahmed, Karine Guillem, and Youna Vandaele, "Sugar Addiction: Pushing the Drug-Sugar Analogy to the Limit," *Current Opinion in Clinical Nutrition & Metabolic Care* 16, no. 4 (2013): 434–439, https://doi.org/10.1097/MCO.0b013e328361c8b8.

6. Barry Reisberg et al., "Clinical Symptoms Accompanying Progressive Cognitive Decline and Alzheimer's Disease," *Alzheimer's Dementia: Dilemmas in Clinical Research* (1985): 19–39.

7. Ramón Estruch et al., "Primary Prevention of Cardiovascular Disease with a Mediterranean Diet," *New England Journal of Medicine* 368, no. 14 (2013): 1279–1290, https://doi.org/10.1056/NEJMoa1200303.

8. Serena Tonstad et al., "Type of Vegetarian Diet, Body Weight, and Prevalence of Type 2 Diabetes," *Diabetes Care* 32, no. 5 (2009): 791–796, https://doi.org/10.2337/dc08-1886.

9. Mary Ann S. Van Duyn and Elizabeth Pivonka, "Overview of the Health Benefits of Fruit and Vegetable Consumption for the Dietetics Professional: Selected Literature," *Journal of the American Dietetic Association* 100, no. 12 (2000): 1511–1521, https://doi.org/10.1016/S0002-8223(00)00420-X.

10. Ambika Satija and Frank B. Hu, "Plant-Based Diets and Cardiovascular Health," *Trends in Cardiovascular Medicine* 28, no. 7 (2018): 437–441, https://doi.org/10.1016/j.tcm.2018.02.004.

11. Yoko Brigitte Wang et al., "The Association between Diet Quality, Plant-Based Diets, Systemic Inflammation, and Mortality Risk: Findings from NHANES," *European Journal of Nutrition* 62, no. 7 (2023): 2723–2737, https://doi.org/10.1007/s00394-023-03191-z.

12. Kyung Hee Lee, Myeounghoon Cha, and Bae Hwan Lee, "Neuroprotective Effect of Antioxidants in the Brain," *International Journal of Molecular Sciences* 21, no. 19 (2020): 7152, https://doi.org/10.3390/ijms21197152.

13. Nour Yahfoufi et al., "The Immunomodulatory and Anti-Inflammatory Role of Polyphenols," *Nutrients* 10, no. 11 (2018): 1618, https://doi.org/10.3390/nu10111618.

14. Giuseppe Caruso et al., "Polyphenols and Neuroprotection: Therapeutic Implications for Cognitive Decline," *Pharmacology & Therapeutics* 232 (2022): 108013, https://doi.org/10.1016/j.pharmthera.2021.108013.

15. Atul Bali and Roopa Naik, "The Impact of a Vegan Diet on Many Aspects of Health: The Overlooked Side of Veganism," *Cureus* 15, no. 2 (2023), https://doi.org/10.7759/cureus.35148.
16. Hideaki Sato et al., "Protein Deficiency-Induced Behavioral Abnormalities and Neurotransmitter Loss in Aged Mice Are Ameliorated by Essential Amino Acids," *Frontiers in Nutrition* 7 (2020): 510349, https://doi.org/10.3389/fnut.2020.00023.
17. Harris R. Lieberman, "Amino Acid and Protein Requirements: Cognitive Performance, Stress and Brain Function," in Committee on Military Nutrition Research, *The Role of Protein and Amino Acids in Sustaining and Enhancing Performance* (National Academy Press, 1999), 289–307.
18. Yoshitaka Kondo et al., "Moderate Protein Intake Percentage in Mice for Maintaining Metabolic Health during Approach to Old Age," *GeroScience* 45, no. 4 (2023): 2707–2726, https://doi.org/10.1007/s11357-023-00797-3.
19. There are many BMI calculators online, including one from the National Heart, Lung, and Blood Institute: https://www.nhlbi.nih.gov/health/educational/lose_wt/BMI/bmicalc.htm.
20. A. David Smith et al., "Homocysteine-Lowering by B Vitamins Slows the Rate of Accelerated Brain Atrophy in Mild Cognitive Impairment: A Randomized Controlled Trial," *PLoS One* 5, no. 9 (2010): e12244, https://doi.org/10.1371/journal.pone.0012244.
21. Nikolaj Travica et al., "Vitamin C Status and Cognitive Function: A Systematic Review," *Nutrients* 9, no. 9 (2017): 960, https://doi.org/10.3390/nu9090960.
22. Liang Shen and Hong-Fang Ji, "Vitamin D Deficiency Is Associated with Increased Risk of Alzheimer's Disease and Dementia: Evidence from Meta-Analysis," *Nutrition Journal* 14 (2015): 1–5, https://doi.org/10.1186/s12937-015-0063-7.
23. Tom C. Russ et al., "Geographical Variation in Dementia Mortality in Italy, New Zealand, and Chile: The Impact of Latitude, Vitamin D, and Air Pollution," *Dementia and Geriatric Cognitive Disorders* 42, no. 1–2 (2016): 31–41, https://doi.org/10.1159/000447449.
24. Maricruz Sepulveda-Villegas, Leticia Elizondo-Montemayor, and Victor Trevino, "Identification and Analysis of 35 Genes Associated with Vitamin D Deficiency: A Systematic Review to Identify Genetic Variants," *Journal of Steroid Biochemistry and Molecular Biology* 196 (2020): 105516, https://doi.org/10.1016/j.jsbmb.2019.105516.
25. Claudia L. Satizabal et al., "Association of Red Blood Cell Omega-3 Fatty Acids with MRI Markers and Cognitive Function in Midlife: The

Framingham Heart Study," *Neurology* 99, no. 23 (2022): e2572–e2582, https://doi.org/10.1212/WNL.0000000000201296.

26. Khawlah Alateeq, Erin I. Walsh, and Nicolas Cherbuin, "Dietary Magnesium Intake Is Related to Larger Brain Volumes and Lower White Matter Lesions with Notable Sex Differences," *European Journal of Nutrition* 62, no. 5 (2023): 2039–2051, https://doi.org/10.1007/s00394-023-03123-x.

27. Maureen M. Black, "The Evidence Linking Zinc Deficiency with Children's Cognitive and Motor Functioning," *Journal of Nutrition* 133, no. 5 (2003): 1473S–1476S, https://doi.org/10.1093/jn/133.5.1473S.

28. Zhe Li et al., "The Important Role of Zinc in Neurological Diseases," *Biomolecules* 13, no. 1 (2022): 28, https://doi.org/10.3390/biom13010028.

29. Cristina Fernández-Portero et al., "Coenzyme Q10 Levels Associated with Cognitive Functioning and Executive Function in Older Adults," *Journals of Gerontology: Series A* 78, no. 1 (2023): 1–8, https://doi.org/10.1093/gerona/glac152.

30. Diana Cardenas, "Let Not Thy Food Be Confused with Thy Medicine: The Hippocratic Misquotation," *e-SPEN Journal* 8, no. 6 (2013): e260–e262, https://doi.org/10.1016/j.clnme.2013.10.002.

31. Martha N. Gardner and Allan M. Brandt, "'The Doctors' Choice Is America's Choice': The Physician in US Cigarette Advertisements, 1930–1953," *American Journal of Public Health* 96, no. 2 (2006): 222–232, https://doi.org/10.2105/AJPH.2005.066654.

32. Erik Peper and Richard Harvey, "Are Food Companies Responsible for the Epidemic in Diabetes, Cancer, Dementia and Chronic Disease and Do Their Products Need to Be Regulated Like Tobacco? Is It Time for a Class Action Suit?," *Townsend Letter*, January 13, 2024, https://www.townsendletter.com/e-letter-26-ultra-processed-foods-and-health-issues/.

33. Natalia Gomes Gonçalves et al., "Association between Consumption of Ultraprocessed Foods and Cognitive Decline," *JAMA Neurology* 80, no. 2 (2023): 142–150, https://doi.org/10.1001/jamaneurol.2022.4397.

34. Virginie Mansuy-Aubert and Yann Ravussin, "Short Chain Fatty Acids: The Messengers from Down Below," *Frontiers in Neuroscience* 17 (2023): 1197759, https://doi.org/10.3389/fnins.2023.1197759.

35. Katie Meyer et al., "Association of the Gut Microbiota with Cognitive Function in Midlife," *JAMA Network Open* 5, no. 2 (2022), https://doi.org/10.1001/jamanetworkopen.2021.43941.

36. Vanessa Ridaura and Yasmine Belkaid, "Gut Microbiota: The Link to Your Second Brain," *Cell* 161, no. 2 (2015): 193–194, https://doi.org/10.1016/j.cell.2015.03.033.

37. Jessica Eastwood et al., "The Effect of Probiotics on Cognitive Function across the Human Lifespan: A Systematic Review," *Neuroscience &*

Biobehavioral Reviews 128 (2021): 311–327, https://doi.org/10.1016/j.neubiorev.2021.06.032.

38. Åsa Hammar and Guro Årdal, "Cognitive Functioning in Major Depression—A Summary," *Frontiers in Human Neuroscience* 3 (2009): 728, https://doi.org/10.3389/neuro.09.026.2009.

39. Robert B. McGandy, D. Mark Hegsted, and F. J. Stare, "Dietary Fats, Carbohydrates and Atherosclerotic Vascular Disease," *New England Journal of Medicine* 277, no. 4 (1967): 186–192, https://www.nejm.org/doi/pdf/10.1056/NEJM196707272770405.

40. Cristin E. Kearns, Laura A. Schmidt, and Stanton A. Glantz, "Sugar Industry and Coronary Heart Disease Research: A Historical Analysis of Internal Industry Documents," *JAMA Internal Medicine* 176, no. 11 (2016): 1680–1685, https://doi.org/10.1001/jamainternmed.2016.5394.

41. Anahad O'Connor, "Coca-Cola Funds Scientists Who Shift Blame for Obesity Away from Bad Diets," *New York Times*, August 9, 2015, https://archive.nytimes.com/well.blogs.nytimes.com/2015/08/09/coca-cola-funds-scientists-who-shift-blame-for-obesity-away-from-bad-diets/.

42. Candice Choi, "How Candy Makers Shape Nutrition Science," Associated Press, June 2, 2016, https://apnews.com/d90190c4a77e470ca0ebd332f3b049fd.

43. David Merritt Johns and Gerald M. Oppenheimer, "Was There Ever Really a 'Sugar Conspiracy'?," *Science* 359, no. 6377 (2018): 747–750, https://doi.org/10.1126/science.aaq1618.

44. Sebastian Brandhorst et al., "Fasting-Mimicking Diet Causes Hepatic and Blood Markers Changes Indicating Reduced Biological Age and Disease Risk," *Nature Communications* 15, no. 1 (2024): 1309, https://doi.org/10.1038/s41467-024-45260-9.

45. Dara L. James et al., "Impact of Intermittent Fasting and/or Caloric Restriction on Aging-Related Outcomes in Adults: A Scoping Review of Randomized Controlled Trials," *Nutrients* 16, no. 2 (2024): 316, https://doi.org/10.3390/nu16020316.

46. Jip Gudden, Alejandro Arias Vasquez, and Mirjam Bloemendaal, "The Effects of Intermittent Fasting on Brain and Cognitive Function," *Nutrients* 13, no. 9 (2021): 3166, https://doi.org/10.3390/nu13093166.

47. Kirrilly M. Pursey et al., "The Prevalence of Food Addiction As Assessed by the Yale Food Addiction Scale: A Systematic Review," *Nutrients* 6, no. 10 (2014): 4552–4590, https://doi.org/10.3390/nu6104552.

48. Amy L. McKenzie and Shaminie J. Athinarayanan, "Impact of Glucagon-Like Peptide 1 Agonist Deprescription in Type 2 Diabetes in a Real-World Setting: A Propensity Score Matched Cohort Study," *Diabetes Therapy* 15, no. 4 (2024): 843–853, https://doi.org/10.1007/s13300-024-01547-0.

8: THE BRAINSPAN WORKOUT

1. J. Wilde-Frenz and H. Schulz, "Rate and Distribution of Body Movements during Sleep in Humans," *Perceptual and Motor Skills* 56, no. 1 (1983): 275–283, https://doi.org/10.2466/pms.1983.56.1.275.
2. Dennis Muñoz-Vergara et al., "Prepandemic Physical Activity and Risk of COVID-19 Diagnosis and Hospitalization in Older Adults," *JAMA Network Open* 7, no. 2 (2024): e2355808, https://doi.org/10.1001/jamanetworkopen.2023.55808.
3. Helen Shinru Wei et al., "Erythrocytes Are Oxygen-Sensing Regulators of the Cerebral Microcirculation," *Neuron* 91, no. 4 (2016): 851–862, https://doi.org/10.1016/j.neuron.2016.07.016.
4. Aaron A. Phillips et al., "Neurovascular Coupling in Humans: Physiology, Methodological Advances and Clinical Implications," *Journal of Cerebral Blood Flow & Metabolism* 36, no. 4 (2016): 647–664, https://doi.org/10.1177/0271678X15617954.
5. Michelle E. Watts, Roger Pocock, and Charles Claudianos, "Brain Energy and Oxygen Metabolism: Emerging Role in Normal Function and Disease," *Frontiers in Molecular Neuroscience* 11 (2018): 216, https://doi.org/10.3389/fnmol.2018.00216.
6. Teresa Liu-Ambrose et al., "Aerobic Exercise and Vascular Cognitive Impairment: A Randomized Controlled Trial," *Neurology* 87, no. 20 (2016): 2082–2090, https://doi.org/10.1212/WNL.0000000000003332.
7. Jun Mu, Paul R. Krafft, and John H. Zhang, "Hyperbaric Oxygen Therapy Promotes Neurogenesis: Where Do We Stand?" *Medical Gas Research* 1 (2011): 1–7, https://doi.org/10.1186/2045-9912-1-14.
8. J. Eric Ahlskog et al., "Physical Exercise As a Preventive or Disease-Modifying Treatment of Dementia and Brain Aging," *Mayo Clinic Proceedings* 86, no. 9 (2011): 876–884, https://doi.org/10.4065/mcp.2011.0252.
9. Mark Evans, Karl E. Cogan, and Brendan Egan, "Metabolism of Ketone Bodies during Exercise and Training: Physiological Basis for Exogenous Supplementation," *Journal of Physiology* 595, no. 9 (2017): 2857–2871, https://doi.org/10.1113/JP273185.
10. Stephanie von Holstein-Rathlou, Nicolas Caesar Petersen, and Maiken Nedergaard, "Voluntary Running Enhances Glymphatic Influx in Awake Behaving, Young Mice," *Neuroscience Letters* 662 (2018): 253–258, https://doi.org/10.1016/j.neulet.2017.10.035.
11. Christopher Ingraham, "Actually, You Do Have Enough Time to Exercise, and Here's the Data to Prove It," *Washington Post*, October 30, 2019, https://www.washingtonpost.com/business/2019/10/30/actually-you-do-have-enough-time-exercise-heres-data-prove-it/.
12. Roland Sturm and Deborah A. Cohen, "Peer Reviewed: Free Time and Phys-

ical Activity among Americans 15 Years or Older: Cross-Sectional Analysis of the American Time Use Survey," *Preventing Chronic Disease* 16 (2019), https://doi.org/10.5888/pcd16.190017.

13. Kathryn M. Broadhouse et al., "Hippocampal Plasticity Underpins Long-Term Cognitive Gains from Resistance Exercise in MCI," *NeuroImage: Clinical* 25 (2020): 102182, https://doi.org/10.1016/j.nicl.2020.102182.

14. Zhihui Li et al., "The Effect of Resistance Training on Cognitive Function in the Older Adults: A Systematic Review of Randomized Clinical Trials," *Aging Clinical and Experimental Research* 30 (2018): 1259–1273, https://doi.org/10.1007/s40520-018-0998-6.

15. McKayla J. Niemann et al., "Strength Training and Insulin Resistance: The Mediating Role of Body Composition," *Journal of Diabetes Research* (2020), https://doi.org/10.1155/2020/7694825.

16. Jorge L. Ruas et al., "A PGC-1α Isoform Induced by Resistance Training Regulates Skeletal Muscle Hypertrophy," *Cell* 151, no. 6 (2012): 1319–1331, https://doi.org/10.1016/j.cell.2012.10.050.

17. Gary Sweeney and Juhyun Song, "The Association between PGC-1α and Alzheimer's Disease," *Anatomy & Cell Biology* 49, no. 1 (2016): 1, https://doi.org/10.5115/acb.2016.49.1.1.

18. Elena Volpi, Reza Nazemi, and Satoshi Fujita, "Muscle Tissue Changes with Aging," *Current Opinion in Clinical Nutrition & Metabolic Care* 7, no. 4 (2004): 405–410, https://doi.org/10.1097/01.mco.0000134362.76653.b2.

19. T. N. Ziegenfuss et al., "Effects of an Amylopectin and Chromium Complex on the Anabolic Response to a Suboptimal Dose of Whey Protein," *Journal of the International Society of Sports Nutrition* 14, no. 1 (2017): 6, https://doi.org/10.53520/jen2021.10394.

20. Gabriel J. Wilson, Jacob M. Wilson, and Anssi H. Manninen, "Effects of Beta-Hydroxy-Beta-Methylbutyrate (HMB) on Exercise Performance and Body Composition across Varying Levels of Age, Sex, and Training Experience: A Review," *Nutrition & Metabolism* 5 (2008): 1–17, https://doi.org/10.1186/1743-7075-5-1.

21. Regina G. Belz and Stephen O. Duke, "Herbicides and Plant Hormesis," *Pest Management Science* 70, no. 5 (2014): 698–707, https://doi.org/10.1002/ps.3726.

22. R. Waziry et al., "Effect of Long-Term Caloric Restriction on DNA Methylation Measures of Biological Aging in Healthy Adults from the CALERIE Trial," *Nature Aging* 3, no. 3 (2023): 248–257, https://doi.org/10.1038/s43587-022-00357-y.

23. David A. Sinclair and Matthew D. LaPlante, *Lifespan: Why We Age—and Why We Don't Have To* (Atria Books, 2019).

24. Matthew M. Robinson et al., "Enhanced Protein Translation Underlies

Improved Metabolic and Physical Adaptations to Different Exercise Training Modes in Young and Old Humans," *Cell Metabolism* 25, no. 3 (2017): 581–592, https://doi.org/10.1016/j.cmet.2017.02.009.

25. Khatija Bahdur et al., "Efecto de HIIT en el rendimiento cognitivo y físico," *Apunts Medicina de l'Esport* 54, no. 204 (2019): 113–117, https://doi.org/10.1016/j.apunts.2019.07.001.

26. Said Mekari et al., "Effect of High Intensity Interval Training Compared to Continuous Training on Cognitive Performance in Young Healthy Adults: A Pilot Study," *Brain Sciences* 10, no. 2 (2020): 81, https://doi.org/10.3390/brainsci10020081.

27. J. P. Loenneke et al., "Blood Flow Restriction Pressure Recommendations: The Hormesis Hypothesis," *Medical Hypotheses* 82, no. 5 (2014): 623–626, https://doi.org/10.1016/j.mehy.2014.02.023.

28. Yudai Takarada et al., "Effects of Resistance Exercise Combined with Moderate Vascular Occlusion on Muscular Function in Humans," *Journal of Applied Physiology* 88, no. 6 (2000): 2097–2106, https://doi.org/10.1152/jappl.2000.88.6.2097.

29. Matthew Futterman, "A Hot Fitness Trend Among Olympians: Blood Flow Restriction," *New York Times*, July 21, 2021.

30. Yudai Takarada et al., "Rapid Increase in Plasma Growth Hormone after Low-Intensity Resistance Exercise with Vascular Occlusion," *Journal of Applied Physiology* 88, no. 1 (2000): 61–65, https://doi.org/10.1152/jappl.2000.88.1.61.

31. Alexander Törpel et al., "Strengthening the Brain—Is Resistance Training with Blood Flow Restriction an Effective Strategy for Cognitive Improvement?," *Journal of Clinical Medicine* 7, no. 10 (2018): 337, https://doi.org/10.3390/jcm7100337.

32. Takeshi Sugimoto et al., "Blood Flow Restriction Improves Executive Function after Walking," *Medicine & Science in Sports & Exercise* 53, no. 1 (2021): 131–138, https://doi.org/10.1249/MSS.0000000000002446.

33. Amir Kargaran et al., "Effects of Dual-Task Training with Blood Flow Restriction on Cognitive Functions, Muscle Quality, and Circulatory Biomarkers in Elderly Women," *Physiology & Behavior* 239 (2021): 113500, https://doi.org/10.1016/j.physbeh.2021.113500.

34. Helen Shinru Wei et al., "Erythrocytes Are Oxygen-Sensing Regulators of the Cerebral Microcirculation," *Neuron* 91, no. 4 (2016): 851–862, https://doi.org/10.1016/j.neuron.2016.07.016.

35. Ling Yan, Ting Liang, and Oumei Cheng, "Hyperbaric Oxygen Therapy in China," *Medical Gas Research* 5 (2015): 1–6, https://doi.org/10.1186/s13618-015-0024-4.

36. Irit Gottfried, Nofar Schottlender, and Uri Ashery, "Hyperbaric Oxygen

Treatment—from Mechanisms to Cognitive Improvement," *Biomolecules* 11, no. 10 (2021): 1520, https://doi.org/10.3390/biom11101520.
37. Enya Daynes et al., "Early Experiences of Rehabilitation for Individuals Post-COVID to Improve Fatigue, Breathlessness Exercise Capacity and Cognition—a Cohort Study," *Chronic Respiratory Disease* 18 (2021): 14799731211015691, https://doi.org/10.1177/14799731211015691.
38. Julie Gregory, "My Experience with EWOT," Apollo Health, March 18, 2024, https://www.apollohealthco.com/my-experience-with-ewot/.
39. Diogo S. Teixeira et al., "Enjoyment As a Predictor of Exercise Habit, Intention to Continue Exercising, and Exercise Frequency: The Intensity Traits Discrepancy Moderation Role," *Frontiers in Psychology* 13 (2022): 780059, https://doi.org/10.3389/fpsyg.2022.780059.
40. Scott M. Lephart et al., "An Eight-Week Golf-Specific Exercise Program Improves Physical Characteristics, Swing Mechanics, and Golf Performance in Recreational Golfers," *Journal of Strength & Conditioning Research* 21, no. 3 (2007): 860–869.
41. Ashley K. Williams, Jonathan Glen, and Graeme G. Sorbie, "The Effect of Upper Body Sprint Interval Training on Golf Drive Performance," *Journal of Sports Medicine and Physical Fitness* 62, no. 11 (2022): 1427–1434, https://doi.org/10.23736/S0022-4707.22.12944-0.
42. Hyun Ahn, Sea-Hyun Bae, and Kyung-Yoon Kim, "Effects of Left Thigh Blood Flow Restriction Exercise on Muscle Strength and Golf Performance in Amateur Golfers," *Journal of Exercise Rehabilitation* 19, no. 4 (2023): 237, https://doi.org/10.12965/jer.2346302.151.
43. Chen-Yu Chen et al., "Early Recovery of Exercise-Related Muscular Injury by HBOT," *BioMed Research International* (2019), https://doi.org/10.1155/2019/6289380.

9: CLEANSE AND RESTORE

1. "Percentage of Adults Aged ≥ 18 Years Who Sleep < 7 Hours on Average in a 24-Hour Period, by Sex and Age Group—National Health Interview Survey, United States, 2020," *Morbidity and Mortality Weekly Report* 71, no. 10 (2022): 393, https://doi.org/10.15585/mmwr.mm7110a6.
2. Caroline Schmidt et al., "Progressive Encephalomyelitis with Rigidity and Myoclonus Preceding Otherwise Asymptomatic Hodgkin's Lymphoma," *Journal of the Neurological Sciences* 291, no. 1–2 (2010): 118–120, https://doi.org/10.1016/j.jns.2009.12.025.
3. Howard S. Kirshner, "Primary Progressive Aphasia and Alzheimer's Disease: Brief History, Recent Evidence," *Current Neurology and Neuroscience Reports* 12 (2012): 709–714, https://doi.org/10.1007/s11910-012-0307-2.

4. Till Roenneberg, "How Can Social Jetlag Affect Health?," *Nature Reviews Endocrinology* 19, no. 7 (2023): 383–384, https://doi.org/10.1038/s41574-023-00851-2.
5. Leah C. Hibel, Evelyn Mercado, and Jill M. Trumbell, "Parenting Stressors and Morning Cortisol in a Sample of Working Mothers," *Journal of Family Psychology* 26, no. 5 (2012): 738, https://doi.org/10.1037/a0029340.
6. Christian Jones et al., "The Impact of Courteous and Discourteous Drivers on Physiological Stress," *Transportation Research Part F: Traffic Psychology and Behaviour* 82 (2021): 285–296, https://doi.org/10.1016/j.trf.2021.08.015.
7. Denise Albieri Jodas Salvagioni et al., "Physical, Psychological and Occupational Consequences of Job Burnout: A Systematic Review of Prospective Studies," *PLoS One* 12, no. 10 (2017): e0185781, https://doi.org/10.1371/journal.pone.0185781.
8. Kjeld Møllgård et al., "A Mesothelium Divides the Subarachnoid Space into Functional Compartments," *Science* 379, no. 6627 (2023): 84–88, https://doi.org/10.1126/science.adc8810.
9. Kenneth I. Hume, Mark Brink, and Mathias Basner, "Effects of Environmental Noise on Sleep," *Noise and Health* 14, no. 61 (2012): 297–302, https://journals.lww.com/nohe/fulltext/2012/14610/effects_of_environmental_noise_on_sleep.6.aspx.
10. Tae-Yoon S. Park et al., "Brain and Eyes of Kerygmachela Reveal Protocerebral Ancestry of the Panarthropod Head," *Nature Communications* 9, no. 1 (2018): 1019, https://doi.org/10.1038/s41467-018-03464-w.
11. Simon Neubauer, Jean-Jacques Hublin, and Philipp Gunz, "The Evolution of Modern Human Brain Shape," *Science Advances* 4, no. 1 (2018): eaao5961, https://doi.org/10.1126/sciadv.aao5961.
12. Richard G. Stevens and Yong Zhu, "Electric Light, Particularly at Night, Disrupts Human Circadian Rhythmicity: Is That a Problem?," *Philosophical Transactions of the Royal Society B: Biological Sciences* 370, no. 1667 (2015): 20140120, https://doi.org/10.1098/rstb.2014.0120.
13. John Marshall, "The Blue Light Paradox: Problem or Panacea," *Points de Vue—International Review of Ophthalmic Optics* (2017).
14. Marin Vogelsang et al., "Prenatal Auditory Experience and Its Sequelae," *Developmental Science* 26, no. 1 (2023): e13278, https://doi.org/10.1111/desc.13278.
15. Thomas Schreiner et al., "Respiration Modulates Sleep Oscillations and Memory Reactivation in Humans," *Nature Communications* 14, no. 1 (2023): 8351, https://doi.org/10.1038/s41467-023-43450-5.
16. Karl A. Franklin and Eva Lindberg, "Obstructive Sleep Apnea Is a Common Disorder in the Population—a Review on the Epidemiology of Sleep Apnea," *Journal of Thoracic Disease* 7, no. 8 (2015): 1311, https://doi.org/10.3978/j.issn.2072-1439.2015.06.11.

Notes

17. Nico J. Diederich et al., "Sleep Apnea Syndrome in Parkinson's Disease. A Case–Control Study in 49 Patients," *Movement Disorders* 20, no. 11 (2005): 1413–1418, https://doi.org/10.1002/mds.20624.
18. Omonigho M. Bubu et al., "Obstructive Sleep Apnea, Cognition and Alzheimer's Disease: A Systematic Review Integrating Three Decades of Multidisciplinary Research," *Sleep Medicine Reviews* 50 (2020): 101250, https://doi.org/10.1016/j.smrv.2019.101250.
19. Wei Xu et al., "Sleep Problems and Risk of All-Cause Cognitive Decline or Dementia: An Updated Systematic Review and Meta-Analysis," *Journal of Neurology, Neurosurgery & Psychiatry* 91, no. 3 (2020): 236–244, https://doi.org/10.1136/jnnp-2019-321896.
20. Kevin K. Motamedi, Andrew C. McClary, and Ronald G. Amedee, "Obstructive Sleep Apnea: A Growing Problem," *Ochsner Journal* 9, no. 3 (2009): 149–153, https://www.ochsnerjournal.org/content/9/3/149.full.
21. Jana R. Cooke et al., "Sustained Use of CPAP Slows Deterioration of Cognition, Sleep, and Mood in Patients with Alzheimer's Disease and Obstructive Sleep Apnea: A Preliminary Study," *Journal of Clinical Sleep Medicine* 5, no. 4 (2009): 305–309, https://doi.org/10.5664/jcsm.27538.
22. Atul Malhotra et al., "Tirzepatide for the Treatment of Obstructive Sleep Apnea and Obesity," *New England Journal of Medicine* (2024), http://doi.org/10.1056/NEJMoa2404881.
23. Yi Xie et al., "Effects of Exercise on Sleep Quality and Insomnia in Adults: A Systematic Review and Meta-Analysis of Randomized Controlled Trials," *Frontiers in Psychiatry* 12 (2021): 664499, https://doi.org/10.3389/fpsyt.2021.664499.
24. Nuria Martinez-Lopez et al., "System-Wide Benefits of Intermeal Fasting by Autophagy," *Cell Metabolism* 26, no. 6 (2017): 856–871, https://doi.org/10.4161/auto.6.6.12376.
25. J. F. Pagel and Bennett L. Parnes, "Medications for the Treatment of Sleep Disorders: An Overview," *Primary Care Companion to the Journal of Clinical Psychiatry* 3, no. 3 (2001): 118, https://doi.org/10.4088/pcc.v03n0303.
26. Christopher N. Kaufmann et al., "Declining Trend in Use of Medications for Sleep Disturbance in the United States from 2013 to 2018," *Journal of Clinical Sleep Medicine* 18, no. 10 (2022): 2459–2465, https://doi.org/10.5664/jcsm.10132.
27. Donald Givler et al., "Chronic Administration of Melatonin: Physiological and Clinical Considerations," *Neurology International* 15, no. 1 (2023): 518–533, https://doi.org/10.3390/neurolint15010031.
28. Anahad O'Connor, "Can Magnesium Supplements Really Help You Sleep?," *New York Times*, August 31, 2021, https://www.nytimes.com/2021/08/31/well/mind/magnesium-supplements-for-sleep.html.

29. Clarinda N. Sutanto, Wen Wei Loh, and Jung Eun Kim, "The Impact of Tryptophan Supplementation on Sleep Quality: A Systematic Review, Meta-Analysis, and Meta-Regression," *Nutrition Reviews* 80, no. 2 (2022): 306–316, https://doi.org/10.1093/nutrit/nuab027.

30. Tayebeh Barsam et al., "Effect of Extremely Low Frequency Electromagnetic Field Exposure on Sleep Quality in High Voltage Substations," *Iranian Journal of Environmental Health Science & Engineering* 9 (2012): 1–7, https://doi.org/10.1186/1735-2746-9-15.

31. E. Díaz-Del Cerro et al., "Improvement of Several Stress Response and Sleep Quality Hormones in Men and Women after Sleeping in a Bed that Protects against Electromagnetic Fields," *Environmental Health* 21, no. 1 (2022): 72, https://doi.org/10.1186/s12940-022-00882-8.

32. Madhav Goyal et al., "Meditation Programs for Psychological Stress and Well-Being: A Systematic Review and Meta-Analysis," *JAMA Internal Medicine* 174, no. 3 (2014): 357–368, https://doi.org/10.1001/jamainternmed.2013.13018.

33. Kathleen C. Spadaro et al., "Effect of Mindfulness Meditation on Short-Term Weight Loss and Eating Behaviors in Overweight and Obese Adults: A Randomized Controlled Trial," *Journal of Complementary and Integrative Medicine* 15, no. 2 (2018), https://doi.org/10.1515/jcim-2016-0048.

34. Heather L. Rusch et al., "The Effect of Mindfulness Meditation on Sleep Quality: A Systematic Review and Meta-Analysis of Randomized Controlled Trials," *Annals of the New York Academy of Sciences* 1445, no. 1 (2019): 5–16, https://doi.org/10.1111/nyas.13996.

35. Tim Gard, Britta K. Hölzel, and Sara W. Lazar, "The Potential Effects of Meditation on Age-Related Cognitive Decline: A Systematic Review," *Annals of the New York Academy of Sciences* 1307, no. 1 (2014): 89–103, https://doi.org/10.1111/nyas.12348.

36. Melanie Curtin, "Neuroscience Reveals 50-Year-Olds Can Have the Brains of 25-Year-Olds If They Do This 1 Thing," *Inc.*, October 23, 2018, https://www.inc.com/melanie-curtin/neuroscience-shows-that-50-year-olds-can-have-brains-of-25-year-olds-if-they-do-this.html.

37. Sara W. Lazar et al., "Meditation Experience Is Associated with Increased Cortical Thickness," *Neuroreport* 16, no. 17 (2005): 1893–1897, https://doi.org/10.1097/01.wnr.0000186598.66243.19.

38. Gretchen A. Brenes et al., "The Effects of Yoga on Patients with Mild Cognitive Impairment and Dementia: A Scoping Review," *American Journal of Geriatric Psychiatry* 27, no. 2 (2019): 188–197, https://doi.org/10.1016/j.jagp.2018.10.013.

39. Peter H. Canter and Edzard Ernst, "The Cumulative Effects of Transcendental Meditation on Cognitive Function—a Systematic Review of Randomised

Controlled Trials," *Wiener Klinische Wochenschrift* 115 (2003): 758–766, https://doi.org/10.1007/BF03040500.
40. Jingjing Yang et al., "Tai Chi Is Effective in Delaying Cognitive Decline in Older Adults with Mild Cognitive Impairment: Evidence from a Systematic Review and Meta-Analysis," *Evidence-Based Complementary and Alternative Medicine* (2020), https://doi.org/10.1155/2020/3620534.
41. Chiew Jiat Rosalind Siah et al., "The Effects of Forest Bathing on Psychological Well-Being: A Systematic Review and Meta-Analysis," *International Journal of Mental Health Nursing* 32, no. 4 (2023): 1038–1054, https://doi.org/10.1111/inm.13131.

10: THE BRAIN'S FLEX FACTOR

1. Kat Toups et al., "Precision Medicine Approach to Alzheimer's Disease: Successful Pilot Project," *Journal of Alzheimer's Disease* 88, no. 4 (2022): 1411–1421, http://doi.org/10.3233/JAD-215707.
2. Heather Sandison et al., "Observed Improvement in Cognition during a Personalized Lifestyle Intervention in People with Cognitive Decline," *Journal of Alzheimer's Disease* 94, no. 3 (2023): 993–1004, https://doi.org/10.3233/JAD-230004.
3. Tara John, "Brain Game App Lumosity to Pay $2 Million Fine for 'Deceptive Advertising,'" *Time*, January 6, 2016, https://time.com/4169123/lumosity-2-million-fine/.
4. Glenn E. Smith et al., "A Cognitive Training Program Based on Principles of Brain Plasticity: Results from the Improvement in Memory with Plasticity-Based Adaptive Cognitive Training (IMPACT) Study," *Journal of the American Geriatrics Society* 57, no. 4 (2009): 594–603, https://doi.org/10.1111/j.1532-5415.2008.02167.x.
5. Dale E. Bredesen, "Reversal of Cognitive Decline: A Novel Therapeutic Program," *Aging* (Albany, NY) 6, no. 9 (2014): 707, https://doi.org/10.18632/aging.100690.
6. Michael M. Merzenich et al., "Progression of Change Following Median Nerve Section in the Cortical Representation of the Hand in Areas 3b and 1 in Adult Owl and Squirrel Monkeys," *Neuroscience* 10, no. 3 (1983): 639–665, https://doi.org/10.1016/0306-4522(83)90208-7.
7. Jerri D. Edwards et al., "Speed of Processing Training Results in Lower Risk of Dementia," *Alzheimer's & Dementia: Translational Research & Clinical Interventions* 3, no. 4 (2017): 603–611, https://doi.org/10.1016/j.trci.2017.09.002.
8. Kat Toups et al., "Precision Medicine Approach to Alzheimer's Disease: Successful Pilot Project," *Journal of Alzheimer's Disease* 88, no. 4 (2022): 1411–1421, https://doi.org/10.3233/JAD-215707.

9. Jia You et al., "40-Hz Rhythmic Visual Stimulation Facilitates Attention by Reshaping the Brain Functional Connectivity," *2020 42nd Annual International Conference of the IEEE Engineering in Medicine & Biology Society* (EMBC), Montreal, QC, Canada, 2873–2876, https://doi.org/10.1109/EMBC44109.2020.9175356.
10. Diane Chan et al., "Gamma Frequency Sensory Stimulation in Mild Probable Alzheimer's Dementia Patients: Results of Feasibility and Pilot Studies," *PLoS One* 17, no. 12 (2022): e0278412, https://doi.org/10.1371/journal.pone.0278412.
11. Derek J. Hoare, Peyman Adjamian, and Magdalena Sereda, "Electrical Stimulation of the Ear, Head, Cranial Nerve, or Cortex for the Treatment of Tinnitus: A Scoping Review," *Neural Plasticity* (2016), https://doi.org/10.1155/2016/5130503.
12. Sophie M.D.D. Fitzsimmons et al., "Repetitive Transcranial Magnetic Stimulation-Induced Neuroplasticity and the Treatment of Psychiatric Disorders: State of the Evidence and Future Opportunities," *Biological Psychiatry* 95, no. 6 (2023): 592–600, https://doi.org/10.1016/j.biopsych.2023.11.016.
13. Débora Buendía et al., "The Transcranial Light Therapy Improves Synaptic Plasticity in the Alzheimer's Disease Mouse Model," *Brain Sciences* 12, no. 10 (2022): 1272, https://doi.org/10.3390/brainsci12101272.
14. Jee Wook Kim et al., "Spouse Bereavement and Brain Pathologies: A Propensity Score Matching Study," *Psychiatry and Clinical Neurosciences* 76, no. 10 (2022): 490–504, https://doi.org/10.1111/pcn.13439.
15. Maria C. Norton et al., "Greater Risk of Dementia When Spouse Has Dementia? The Cache County Study," *Journal of the American Geriatrics Society* 58, no. 5 (2010): 895–900, https://doi.org/10.1111/j.1532-5415.2010.02806.x.
16. Yuan Chang Leong et al., "Conservative and Liberal Attitudes Drive Polarized Neural Responses to Political Content," *Proceedings of the National Academy of Sciences* 117, no. 44 (2020): 27731–27739, https://doi.org/10.1073/pnas.200853011.
17. Noreena Hertz, *The Lonely Century: How to Restore Human Connection in a World That's Pulling Apart*, (Crown Currency, 2021).
18. Eunju Jin and Samuel Suk-Hyun Hwang, "A Preliminary Study on the Neurocognitive Deficits Associated with Loneliness in Young Adults," *Frontiers in Public Health* 12 (2024): 1371063, https://doi.org/10.3389/fpubh.2024.1371063.
19. Jennie Allen, *Find Your People: Building Deep Community in a Lonely World* (WaterBrook, 2022).
20. Marisa G. Franco, *Platonic: How the Science of Attachment Can Help You Make—and Keep—Friends* (Penguin, 2022).

Notes

21. Andy Field, *Encounterism: The Neglected Joys of Being in Person* (W. W. Norton, 2023).
22. Oscar Ybarra et al., "Friends (and Sometimes Enemies) with Cognitive Benefits: What Types of Social Interactions Boost Executive Functioning?," *Social Psychological and Personality Science* 2, no. 3 (2011): 253–261, https://doi.org/10.1177/19485506103868.
23. Jamie Waters, "'The Assignment Made Me Gulp': Could Talking to Strangers Change My Life?," *Guardian,* July 31, 2021, https://www.theguardian.com/lifeandstyle/2021/jul/31/the-assignment-made-me-gulp-could-talking-to-strangers-change-my-life.
24. Robert Plutchik, "The Nature of Emotions: Human Emotions Have Deep Evolutionary Roots, a Fact That May Explain Their Complexity and Provide Tools for Clinical Practice," *American Scientist* 89, no. 4 (2001): 344–350, https://www.jstor.org/stable/27857503.
25. Holly Elser et al., "Association of Early-, Middle-, and Late-Life Depression with Incident Dementia in a Danish Cohort," *JAMA Neurology* 80, no. 9 (2023): 949–958, https://doi.org/10.1001/jamaneurol.2023.2309.
26. Joe Curran et al., "How Does Therapy Harm? A Model of Adverse Process Using Task Analysis in the Meta-Synthesis of Service Users' Experience," *Frontiers in Psychology* 10 (2019): 347, https://doi.org/10.3389/fpsyg.2019.00347.

11: THE TOXIC ADVENTURE

1. Ritchie C. Shoemaker, Dennis House, and James C. Ryan, "Defining the Neurotoxin Derived Illness Chronic Ciguatera Using Markers of Chronic Systemic Inflammatory Disturbances: A Case/Control Study," *Neurotoxicology and Teratology* 32, no. 6 (2010): 633–639, https://doi.org/10.1016/j.ntt.2010.05.007.
2. Dale E. Bredesen, "Inhalational Alzheimer's Disease: An Unrecognized—and Treatable—Epidemic," *Aging* (Albany, NY) 8, no. 2 (2016): 304, https://doi.org/10.18632/aging.100896.
3. Jotin Gogoi et al., "Switching a Conflicted Bacterial DTD-tRNA Code Is Essential for the Emergence of Mitochondria," *Science Advances* 8, no. 2 (2022): eabj7307, https://doi.org/10.1126/sciadv.abj7307.
4. Hussein S. Hussein and Jeffrey M. Brasel, "Toxicity, Metabolism, and Impact of Mycotoxins on Humans and Animals," *Toxicology* 167, no. 2 (2001): 101–134, https://doi.org/10.1016/S0300-483X(01)00471-1.
5. J. W. Bennett and M. Klich, "Mycotoxins," *Clinical Microbiology Reviews* 16, no. 3 (2003): 497, https://doi.org/10.1128/CMR.16.3.497–516.2003.
6. Diana Pisa et al., "Different Brain Regions Are Infected with Fungi in Alzheimer's Disease," *Scientific Reports* 5, no. 1 (2015): 1–13, https://doi.org/10.1038/srep15015.

7. Xi-Dai Long et al. "Polymorphisms in the Coding Region of X-Ray Repair Complementing Group 4 and Aflatoxin B_1-Related Hepatocellular Carcinoma," *Hepatology* 58, no. 1 (2013): 171–181, https://doi.org/10.1002/hep.26311.
8. Rachel Morello-Frosch et al., "Environmental Chemicals in an Urban Population of Pregnant Women and Their Newborns from San Francisco," *Environmental Science & Technology* 50, no. 22 (2016): 12464–12472, https://doi.org/10.1021/acs.est.6b03492.
9. Matthew D. LaPlante, "Families Blame Illnesses on Mold in Their Hill Air Force Base Housing," *Salt Lake Tribune*, April 20, 2007, https://archive.sltrib.com/story.php?ref=/news/ci_5710890.
10. John M. Donnelly, "Members Irate As Some Military Tenants' Rights Ignored," *Roll Call*, Oct. 6, 2022, https://rollcall.com/2022/10/06/members-irate-as-some-military-tenants-rights-ignored/.
11. Maryam Vasefi et al., "Environmental Toxins and Alzheimer's Disease Progression," *Neurochemistry International* 141 (2020): 104852, https://doi.org/10.1016/j.neuint.2020.104852.
12. Jovana Kos et al. "Climate Change and Mycotoxins Trends in Serbia and Croatia: A 15-Year Review," *Foods* 13, no. 9 (2024): 1391, https://doi.org/10.3390/foods13091391.
13. Maja Peraica, Darko Richter, and Dubravka Rašić, "Mycotoxicoses in Children." *Archives of Industrial Hygiene and Toxicology* 65, no. 4 (2014): 347–363. https://doi.org/10.2478/10004-1254-65-2014-2557.
14. Wieslaw Jedrychowski et al., "Cognitive Function of 6-Year-Old Children Exposed to Mold-Contaminated Homes in Early Postnatal Period. Prospective Birth Cohort Study in Poland," *Physiology & Behavior* 104, no. 5 (2011): 989–995, https://doi.org/10.1016/j.physbeh.2011.06.019.
15. Barbara De Santis et al., "Role of Mycotoxins in the Pathobiology of Autism: A First Evidence," *Nutritional Neuroscience* 22, no. 2 (2019): 132–144, https://doi.org/10.1080/1028415X.2017.1357793.
16. Dylan Goetz, "Flint Lost 20% of Its Population in Past Decade, Census Data Shows," *MLive*, August 13, 2021, https://www.mlive.com/news/flint/2021/08/flint-lost-20-of-its-population-in-past-decade-census-data-shows.html.
17. Dale E. Bredesen et al., "Reversal of Cognitive Decline: 100 Patients," *Journal of Alzheimer's Disease and Parkinsonism* 8, no. 450 (2018): 2161–0460, https://doi.org/10.4172/2161-0460.1000450.
18. Kelly M. Bakulski et al., "Heavy Metals Exposure and Alzheimer's Disease and Related Dementias," *Journal of Alzheimer's Disease* 76, no. 4 (2020): 1215–1242, https://doi.org/10.3233/JAD-200282.
19. Allison Kite, "'Time Bomb' Lead Pipes Will Be Removed. But First Wa-

ter Utilities Have to Find Them," *NPR*, July 20, 2022, https://www.npr.org/sections/health-shots/2022/07/20/1112049811/lead-pipe-removal.

20. T. Maphanga, et al., "The Interplay between Temporal and Seasonal Distribution of Heavy Metals and Physiochemical Properties in Kaap River," *International Journal of Environmental Science and Technology* 21 (2024): 6053–6064, https://doi.org/10.1007/s13762-023-05401-x.

21. Zhenzhong Zeng et al., "A Reversal in Global Terrestrial Stilling and Its Implications for Wind Energy Production," *Nature Climate Change* 9, no. 12 (2019): 979–985, https://doi.org/10.1038/s41558-019-0622-6.

22. Lilian Calderón-Garcidueñas et al., "Metals, Nanoparticles, Particulate Matter, and Cognitive Decline," *Frontiers in Neurology* 12 (2022): 794071, https://doi.org/10.3389/fneur.2021.794071.

23. John K. Kodros et al., "Quantifying the Health Benefits of Face Masks and Respirators to Mitigate Exposure to Severe Air Pollution," *GeoHealth* 5, no. 9 (2021): e2021GH000482, https://doi.org/10.1029/2021GH000482.

24. Francine K. Welty, "Omega-3 Fatty Acids and Cognitive Function," *Current Opinion in Lipidology* 34, no. 1 (2023): 12–21, https://doi.org/10.1097/MOL.0000000000000862.

25. Diana Echeverria et al., "Chronic Low-Level Mercury Exposure, BDNF Polymorphism, and Associations with Cognitive and Motor Function," *Neurotoxicology and Teratology* 27, no. 6 (2005): 781–796, https://doi.org/10.1016/j.ntt.2005.08.001.

26. Madeleine H. Milne et al., "Exposure of U.S. Adults to Microplastics from Commonly-Consumed Proteins," *Environmental Pollution* 343 (2024): 123233, https://doi.org/10.1016/j.envpol.2023.123233.

27. Karina Huynh, "Presence of Microplastics in Carotid Plaques Linked to Cardiovascular Events," *Nature Reviews Cardiology* 21, no. 5 (2024): 279, https://doi.org/10.1038/s41569-024-01015-z.

28. Lauren C. Jenner et al., "Detection of Microplastics in Human Lung Tissue Using μFTIR Spectroscopy," *Science of the Total Environment* 831 (2022): 154907, https://doi.org/10.1016/j.scitotenv.2022.154907.

29. Marcus M. Garcia et al., "In Vivo Tissue Distribution of Polystyrene or Mixed Polymer Microspheres and Metabolomic Analysis after Oral Exposure in Mice," *Environmental Health Perspectives* 132, no. 4 (2024): 047005, https://doi.org/10.1289/EHP13435.

30. Lauren Gaspar et al., "Acute Exposure to Microplastics Induced Changes in Behavior and Inflammation in Young and Old Mice," *International Journal of Molecular Sciences* 24, no. 15 (2023): 12308, https://doi.org/10.3390/ijms241512308.

31. Romilly E. Hodges and Deanna M. Minich, "Modulation of Metabolic Detoxification Pathways Using Foods and Food-Derived Components: A Scientific Review with Clinical Application," *Journal of Nutrition and Metabolism* 2015, no. 1 (2015): 760689, https://doi.org/10.1155/2015/760689.

32. Swathi Suresh, Ankul Singh, and Chitra Vellapandian, "Bisphenol A Exposure Links to Exacerbation of Memory and Cognitive Impairment: A Systematic Review of the Literature," *Neuroscience & Biobehavioral Reviews* 143 (2022): 104939, https://doi.org/10.1016/j.neubiorev.2022.104939.

33. Stephen J. Genuis et al., "Human Excretion of Bisphenol A: Blood, Urine, and Sweat (BUS) Study," *Journal of Environmental and Public Health* 2012, no. 1 (2012): 185731, https://doi.org/10.1155/2012/185731.

34. Nikki L. Hill et al., "Patient-Provider Communication about Cognition and the Role of Memory Concerns: A Descriptive Study," *BMC Geriatrics* 23, no. 1 (2023): 342, https://doi.org/10.1186/s12877-023-04053-3.

35. Douglas K. Owens et al., "Screening for Cognitive Impairment in Older Adults: U.S. Preventive Services Task Force Recommendation Statement," *JAMA* 323, no. 8 (2020): 757–763, https://doi.org/10.1001/jama.2020.0435.

36. Hannah T. Neprash et al., "Association of Primary Care Visit Length with Potentially Inappropriate Prescribing," *JAMA Health Forum* 4, no. 3 (2023): e230052, https://doi.org/10.1001/jamahealthforum.2023.0052.

37. Vipin Soni et al., "Effects of VOCs on Human Health," in *Air Pollution and Control*, ed. Nikhil Sharma, Avinash Kumar Agarwal, Peter Eastwood, Tarun Gupta, and Akhilendra P. Singh (Springer, Signapore, 2018), 119–142, https://doi.org/10.1007/978-981-10-7185-0.

38. Jason Kilian and Masashi Kitazawa, "The Emerging Risk of Exposure to Air Pollution on Cognitive Decline and Alzheimer's Disease—Evidence from Epidemiological and Animal Studies," *Biomedical Journal* 41, no. 3 (2018): 141–162, https://doi.org/10.1016/j.bj.2018.06.001.

39. Bin Jiao et al., "A Detection Model for Cognitive Dysfunction Based on Volatile Organic Compounds from a Large Chinese Community Cohort," *Alzheimer's & Dementia* 19, no. 11 (2023): 4852–4862, https://doi.org/10.1002/alz.13053.

40. Lisa L. von Moltke et al., "Cognitive Toxicity of Drugs Used in the Elderly," *Dialogues in Clinical Neuroscience* 3, no. 3 (2001): 181–190, https://doi.org/10.31887/DCNS.2001.3.3/llvonmoltke.

41. Joy N. Hussain, Ronda F. Greaves, and Marc M. Cohen, "A Hot Topic for Health: Results of the Global Sauna Survey," *Complementary Therapies in Medicine* 44 (2019): 223–234, https://doi.org/10.1016/j.ctim.2019.03.012.

42. Paul Knekt et al., "Does Sauna Bathing Protect against Dementia?," *Preven-

tive Medicine Reports 20 (2020): 101221, https://doi.org/10.1016/j.pmedr.2020.101221.

43. Yigal Erel et al., "Lead in Archeological Human Bones Reflecting Historical Changes in Lead Production," *Environmental Science & Technology* 55, no. 21 (2021): 14407–14413, https://doi.org/10.1021/acs.est.1c00614.

44. Stephanie Than et al., "Cognitive Trajectories during the Menopausal Transition," *Frontiers in Dementia* 2 (2023): 1098693, https://doi.org/10.3389/frdem.2023.1098693.

45. Margit L. Bleecker et al., "The Association of Lead Exposure and Motor Performance Mediated by Cerebral White Matter Change," *Neurotoxicology* 28, no. 2 (2007): 318–323, https://doi.org/10.1016/j.neuro.2006.04.008.

12: THE MICROBIAL MIND

1. Kari E. Murros et al., "*Desulfovibrio* Bacteria Are Associated with Parkinson's Disease," *Frontiers in Cellular and Infection Microbiology* 11 (2021): 652617, https://doi.org/10.3389/fcimb.2021.652617.

2. Sudha B. Singh, Amanda Carroll-Portillo, and Henry C. Lin, "*Desulfovibrio* in the Gut: The Enemy Within?," *Microorganisms* 11, no. 7 (2023): 1772, https://doi.org/10.3390/microorganisms11071772.

3. Vy A. Huynh et al., "*Desulfovibrio* Bacteria Enhance Alpha-Synuclein Aggregation in a *Caenorhabditis Elegans* Model of Parkinson's Disease," *Frontiers in Cellular and Infection Microbiology* 13 (2023): 1181315, https://doi.org/10.3389/fcimb.2023.1181315.

4. Friedrich Leblhuber et al., "The Immunopathogenesis of Alzheimer's Disease Is Related to the Composition of Gut Microbiota," *Nutrients* 13, no. 2 (2021): 361, https://doi.org/10.3390/nu13020361.

5. Selma P. Wiertsema et al., "The Interplay between the Gut Microbiome and the Immune System in the Context of Infectious Diseases throughout Life and the Role of Nutrition in Optimizing Treatment Strategies," *Nutrients* 13, no. 3 (2021): 886, https://doi.org/10.3390/nu13030886.

6. Anne Maczulak, *Allies and Enemies: How the World Depends on Bacteria* (FT Press, 2010).

7. Tatsuya Dokoshi et al., "Dermal Injury Drives a Skin to Gut Axis That Disrupts the Intestinal Microbiome and Intestinal Immune Homeostasis in Mice," *Nature Communications* 15, no. 1 (2024): 3009, https://doi.org/10.1038/s41467-024-47072-3.

8. Katie Meyer et al., "Association of the Gut Microbiota with Cognitive Function in Midlife," *JAMA Network Open* 5, no. 2 (2022): e2143941, https://doi.org/10.1001/jamanetworkopen.2021.43941.

9. Jessica Eastwood et al., "The Effect of Probiotics on Cognitive Function across the Human Lifespan: A Systematic Review," *Neuroscience & Biobehavioral Reviews* 128 (2021): 311–327, https://doi.org/10.1016/j.neubiorev.2021.06.032.

10. Weiai Jia et al., "Association between Dietary Vitamin B$_1$ Intake and Cognitive Function among Older Adults: A Cross-Sectional Study," *Journal of Translational Medicine* 22, no. 1 (2024): 165, https://doi.org/10.1186/s12967-024-04969-3.

11. Mengran Zhang et al., "Biomimetic Remodeling of Microglial Riboflavin Metabolism Ameliorates Cognitive Impairment by Modulating Neuroinflammation," *Advanced Science* 10, no. 12 (2023): 2300180, https://doi.org/10.1002/advs.202300180.

12. Kai Zhang et al., "Association between Dietary Niacin Intake and Cognitive Impairment in Elderly People: A Cross-Sectional Study," *European Journal of Psychiatry* 38, no. 3 (2024): 100233, https://doi.org/10.1016/j.ejpsy.2023.100233.

13. Jingshu Xu et al., "Cerebral Deficiency of Vitamin B5 (d-pantothenic acid; pantothenate) As a Potentially-Reversible Cause of Neurodegeneration and Dementia in Sporadic Alzheimer's Disease," *Biochemical and Biophysical Research Communications* 527, no. 3 (2020): 676–681, https://doi.org/10.1016/j.bbrc.2020.05.015.

14. Hui Xu et al., "Vitamin B$_6$, B$_9$, and B$_{12}$ Intakes and Cognitive Performance in Elders: National Health and Nutrition Examination Survey, 2011–2014." *Neuropsychiatric Disease and Treatment* 18 (2022): 537–553, https://doi.org/10.2147/NDT.S337617.

15. Athena Enderami, Mehran Zarghami, and Hadi Darvishi-Khezri, "The Effects and Potential Mechanisms of Folic Acid on Cognitive Function: A Comprehensive Review," *Neurological Sciences* 39 (2018): 1667–1675, https://doi.org/10.1007/s10072-018-3473-4.

16. Shazia Jatoi et al., "Low Vitamin B$_{12}$ Levels: An Underestimated Cause of Minimal Cognitive Impairment and Dementia," *Cureus* 12, no. 2 (2020), https://doi.org/10.7759/cureus.6976.

17. Ying Zhou et al., "Folate and Vitamin B$_{12}$ Usual Intake and Biomarker Status by Intake Source in United States Adults Aged ≥ 19 y: NHANES 2007–2018," *American Journal of Clinical Nutrition* 118, no. 1 (2023): 241–254, https://doi.org/10.1016/j.ajcnut.2023.05.016.

18. Ludovico Alisi et al., "The Relationships between Vitamin K and Cognition: A Review of Current Evidence," *Frontiers in Neurology* 10 (2019): 416803, https://doi.org/10.3389/fneur.2019.00239.

19. Stephen S. Dominy et al., "*Porphyromonas Gingivalis* in Alzheimer's Disease

20. Saori Nonaka, Tomoko Kadowaki, and Hiroshi Nakanishi, "Secreted Gingipains from *Porphyromonas Gingivalis* Increase Permeability in Human Cerebral Microvascular Endothelial Cells through Intracellular Degradation of Tight Junction Proteins," *Neurochemistry International* 154 (2022): 105282, https://doi.org/10.1016/j.neuint.2022.105282.
21. Lindsey Wang et al., "Association of COVID-19 with New-Onset Alzheimer's Disease," *Journal of Alzheimer's Disease* 89, no. 2 (2022): 411–414, https://doi.org/10.3233/JAD-220717.
22. Christian Zanza et al., "Cytokine Storm in COVID-19: Immunopathogenesis and Therapy," *Medicina* 58, no. 2 (2022): 144, https://doi.org/10.3390/medicina58020144.
23. Elisa Gouvea Gutman et al., "Long COVID: Plasma Levels of Neurofilament Light Chain in Mild COVID-19 Patients with Neurocognitive Symptoms," *Molecular Psychiatry* 29 (2024): 3106–3116, https://doi.org/10.1038/s41380-024-02554-0.
24. Steven Lehrer and Peter H. Rheinstein, "Vaccination Reduces Risk of Alzheimer's Disease, Parkinson's Disease, and Other Neurodegenerative Disorders," *Discovery Medicine* 34, no. 172 (2022): 97–101, https://www.discoverymedicine.com/Steven-Lehrer/2022/10/vaccination-reduces-risk-alzheimers-disease-parkinsons-disease-neurodegeneration/.
25. Jeffrey F. Scherrer et al., "Lower Risk for Dementia Following Adult Tetanus, Diphtheria, and Pertussis (Tdap) Vaccination," *Journals of Gerontology: Series A* 76, no. 8 (2021): 1436–1443, https://doi.org/110.1093/gerona/glab115.
26. Jee Hoon Roh et al., "A Potential Association between COVID-19 Vaccination and Development of Alzheimer's Disease," *QJM: An International Journal of Medicine* (2024): hcae103, https://doi.org/10.1093/qjmed/hcae103.
27. Muazzam Nasrullah et al., "Factors Associated with Condom Use among Sexually Active U.S. Adults, National Survey of Family Growth, 2006–2010 and 2011–2013," *Journal of Sexual Medicine* 14, no. 4 (2017): 541–550, https://doi.org/10.1016/j.jsxm.2017.02.015.
28. Kim Tingley, "Why Are Sexually Transmitted Infections Surging?," *New York Times Magazine*, May 17, 2022, https://www.nytimes.com/2022/05/17/magazine/sexually-transmitted-infections-surging.html.
29. Avinash Rao et al., "Neurosyphilis: An Uncommon Cause of Dementia," *Journal of the American Geriatrics Society* 63, no. 8 (2015), https://doi.org/10.1111/jgs.13571.

30. Ruth F. Itzhaki et al., "Microbes and Alzheimer's Disease," *Journal of Alzheimer's Disease* 51, no. 4 (2016): 979–984, https://doi.org/10.3233/JAD-160152.
31. Katharine S. Walter et al., "Genomic Insights into the Ancient Spread of Lyme Disease across North America," *Nature Ecology & Evolution* 1, no. 10 (2017): 1569–1576, https://doi.org/10.1038/s41559-017-0282-8.
32. Germán Bersalli, Tim Tröndle, and Johan Lilliestam, "Most Industrialised Countries Have Peaked Carbon Dioxide Emissions during Economic Crises through Strengthened Structural Change," *Communications Earth & Environment* 4, no. 1 (2023): 44, https://doi.org/10.1038/s43247-023-00687-8.
33. M. T. Dvorak et al., "Estimating the Timing of Geophysical Commitment to 1.5 and 2.0 C of Global Warming," *Nature Climate Change* 12, no. 6 (2022): 547–552, https://doi.org/10.1038/s41558-022-01372-y.
34. Jennifer M. Coughlin et al., "Imaging Glial Activation in Patients with Post-Treatment Lyme Disease Symptoms: A Pilot Study Using [11 C] DPA-713 PET," *Journal of Neuroinflammation* 15 (2018): 1–7, https://doi.org/10.1186/s12974-018-1381-4.
35. Sumadhya D. Fernando, Chaturaka Rodrigo, and Senaka Rajapakse, "The 'Hidden' Burden of Malaria: Cognitive Impairment Following Infection," *Malaria Journal* 9 (2010): 336, https://doi.org/10.1186/1475-2875-9-366.
36. Apoorva Mandavilli, "At Long Last, Can Malaria Be Eradicated?," *New York Times*, October 4, 2022, https://www.nytimes.com/2022/10/04/health/malaria-vaccines.html.
37. "Editorial: Microbiology by Numbers," *Nature Reviews Microbiology* 9, no. 628 (2011), https://doi.org/10.1038/nrmicro2644.
38. Rino Rappuoli et al., "Save the Microbes to Save the Planet. A Call to Action of the International Union of the Microbiological Societies (IUMS)," *One Health Outlook* 5, no. 1 (2023): 5, https://doi.org/10.1186/s42522-023-00077-2.
39. David A. Sinclair and Matthew D. LaPlante, *Lifespan: Why We Age—and Why We Don't Have To* (Atria Books, 2019).

13: SOONISM

1. Jared M. Campbell, "Supplementation with NAD+ and Its Precursors to Prevent Cognitive Decline across Disease Contexts," *Nutrients* 14, no. 15 (2022): 3231, https://doi.org/10.3390/nu14153231.
2. Matilde Nerattini et al., "Systematic Review and Meta-Analysis of the Effects of Menopause Hormone Therapy on Risk of Alzheimer's Disease and Dementia," *Frontiers in Aging Neuroscience* 15 (2023): 1260427, https://doi.org/10.3389/fnagi.2023.1260427.
3. Yu Jin Kim et al., "Menopausal Hormone Therapy and Risk of Neurodegen-

erative Diseases: Implications for Precision Hormone Therapy," *Alzheimer's & Dementia: Translational Research & Clinical Interventions* 7, no. 1 (2021): e12174, https://doi.org/10.1002/trc2.12174.

4. Karen K. Ryan and Randy J. Seeley, "Food As a Hormone," *Science* 339, no. 6122 (2013): 918–919, https://doi.org/10.1126/science.123406.

5. Anthony C. Hackney and Amy R. Lane, "Exercise and the Regulation of Endocrine Hormones," *Progress in Molecular Biology and Translational Science* 135 (2015): 293–311, https://doi.org/10.1016/bs.pmbts.2015.07.001.

6. Dun-Xian Tan, "Editorial [Hot Topic: Melatonin and Brain (Guest Editor: Dun-Xian Tan)]," *Current Neuropharmacology* 8, no. 3 (2010): 161, https://doi.org/10.2174/157015910792246263.

7. Dewan Md Sumsuzzman et al., "Neurocognitive Effects of Melatonin Treatment in Healthy Adults and Individuals with Alzheimer's Disease and Insomnia: A Systematic Review and Meta-Analysis of Randomized Controlled Trials," *Neuroscience & Biobehavioral Reviews* 127 (2021): 459–473, https://doi.org/10.1016/j.neubiorev.2021.04.034.

8. Laura D. Baker et al., "Effects of Growth Hormone–Releasing Hormone on Cognitive Function in Adults with Mild Cognitive Impairment and Healthy Older Adults: Results of a Controlled Trial," *Archives of Neurology* 69, no. 11 (2012): 1420–1429, https://doi.org/10.1001/archneurol.2012.1970.

9. Bu B. Yeap and Leon Flicker, "Testosterone, Cognitive Decline and Dementia in Ageing Men," *Reviews in Endocrine and Metabolic Disorders* 23, no. 6 (2022): 1243–1257, https://doi.org/10.1007/s11154-022-09728-7.

10. Trey Sunderland et al., "Reduced Plasma Dehydroepiandrosterone Concentrations in Alzheimer's Disease," *Lancet* 334, no. 8662 (1989): 570, https://doi.org/10.1016/S0140-6736(89)90700-9.

11. Yi Zhu et al, "Orally-Active, Clinically-Translatable Senolytics Restore α-Klotho in Mice and Humans," *Lancet* 77 (2022), https://www.thelancet.com/journals/ebiom/article/PIIS2352-3964(22)00096-2/fulltext.

12. Xiaojin Sun et al., "AngIV-Analog Dihexa Rescues Cognitive Impairment and Recovers Memory in the APP/PS1 Mouse via the PI3K/AKT Signaling Pathway," *Brain Sciences* 11, no. 11 (2021): 1487, https://doi.org/10.3390/brainsci11111487.

13. Alaeddine Djillani et al., "Shortened Spadin Analogs Display Better TREK-1 Inhibition, *In Vivo* Stability and Antidepressant Activity," *Frontiers in Pharmacology* 8 (2017): 643, https://doi.org/10.3389/fphar.2017.00643.

14. G. V. Karantysh et al., "Effect of Pinealon on Learning and Expression of NMDA Receptor Subunit Genes in the Hippocampus of Rats with Experimental Diabetes," *Neurochemical Journal* 14, no. 3 (2020): 314–320, https://doi.org/10.1134/S181971242003006X.

15. Tetsuo Shoji et al., "Clinical Availability of Serum Fructosamine Measurement in Diabetic Patients with Uremia: Use As a Glycemic Index in Uremic Diabetes," *Nephron* 51, no. 3 (1989): 338–343, https://doi.org/10.1159/000185319.
16. Asimina Dominari et al., "Thymosin Alpha 1: A Comprehensive Review of the Literature," *World Journal of Virology* 9, no. 5 (2020): 67, http://doi.org/10.5501/wjv.v9.i5.67.
17. Ye Xiong et al., "Neuroprotective and Neurorestorative Effects of Thymosin β4 Treatment Following Experimental Traumatic Brain Injury," *Annals of the New York Academy of Sciences* 1270, no. 1 (2012): 51–58, https://doi.org/10.1111/j.1749-6632.2012.06683.x.
18. Adam L. Boxer et al., "Davunetide in Patients with Progressive Supranuclear Palsy: A Randomised, Double-Blind, Placebo-Controlled Phase 2/3 Trial," *Lancet Neurology* 13, no. 7 (2014): 676–685, https://doi.org/10.1016/S1474-4422(14)70088-2.
19. Dale E. Bredesen, "Reversal of Cognitive Decline: A Novel Therapeutic Program," *Aging* (Albany, NY) 6, no. 9 (2014): 707, https://doi.org/10.18632/aging.100690.
20. Bin Xue et al., "Brain-Derived Neurotrophic Factor: A Connecting Link Between Nutrition, Lifestyle, and Alzheimer's Disease," *Frontiers in Neuroscience* 16 (2022): 925991, https://doi.org/10.3389/fnins.2022.925991.
21. Fernando Gómez-Pinilla et al., "Voluntary Exercise Induces a BDNF-Mediated Mechanism That Promotes Neuroplasticity," *Journal of Neurophysiology* 88, no. 5 (2002): 2187–2195, https://doi.org/10.1152/jn.00152.2002.
22. Chun Chen et al., "Optimized TrkB Agonist Ameliorates Alzheimer's Disease Pathologies and Improves Cognitive Functions via Inhibiting Delta-Secretase," *ACS Chemical Neuroscience* 12, no. 13 (2021): 2448–2461, https://doi.org/10.1021/acschemneuro.1c00181.
23. Jennifer L. Robinson et al., "Neurophysiological Effects of Whole Coffee Cherry Extract in Older Adults with Subjective Cognitive Impairment: A Randomized, Double-Blind, Placebo-Controlled, Cross-Over Pilot Study," *Antioxidants* 10, no. 2 (2021): 144, https://doi.org/10.3390/antiox10020144.
24. Helga Eyjolfsdottir et al., "Targeted Delivery of Nerve Growth Factor to the Cholinergic Basal Forebrain of Alzheimer's Disease Patients: Application of a Second-Generation Encapsulated Cell Biodelivery Device," *Alzheimer's Research & Therapy* 8 (2016): 30, https://doi.org/10.1186/s13195-016-0195-9.
25. Chang-Hun Chae and Hyun-Tae Kim, "Forced, Moderate-Intensity Treadmill Exercise Suppresses Apoptosis by Increasing the Level of NGF and Stimulating Phosphatidylinositol 3-Kinase Signaling in the Hippocampus of

Induced Aging Rats," *Neurochemistry International* 55, no. 4 (2009): 208–213, https://doi.org/10.1016/j.neuint.2009.02.024.

26. Thulasi Ramani et al., "Cytokines: The Good, the Bad, and the Deadly," *International Journal of Toxicology* 34, no. 4 (2015): 355–365, https://doi.org/10.1177/1091581815584918.

27. Andrew J. Westwood et al., "Insulin-Like Growth Factor-1 and Risk of Alzheimer Dementia and Brain Atrophy," *Neurology* 82, no. 18 (2014): 1613–1619, https://doi.org/10.1212/WNL.0000000000000382.

28. William B. Zhang et al., "The Antagonistic Pleiotropy of Insulin-Like Growth Factor 1," *Aging Cell* 20, no. 9 (2021): e13443, https://doi.org/10.1111/acel.13443.

29. Zhi Cao et al., "Circulating Insulin-Like Growth Factor-1 and Brain Health: Evidence from 369,711 Participants in the UK Biobank," *Alzheimer's Research & Therapy* 15, no. 1 (2023): 140, https://doi.org/10.1186/s13195-023-01288-5.

30. Sriram Gubbi et al., "40 YEARS of IGF1: IGF1: The Jekyll and Hyde of the Aging Brain," *Journal of Molecular Endocrinology* 61, no. 1 (2018): T171–T185, https://doi.org/10.1530/JME-18-0093.

31. Asma Kazemi et al., "Effect of Calorie Restriction or Protein Intake on Circulating Levels of Insulin Like Growth Factor I in Humans: A Systematic Review and Meta-Analysis," *Clinical Nutrition* 39, no. 6 (2020): 1705–1716, https://doi.org/10.1016/j.clnu.2019.07.030.

32. Sumonto Mitra et al., "Increased Endogenous GDNF in Mice Protects against Age-Related Decline in Neuronal Cholinergic Markers," *Frontiers in Aging Neuroscience* 13 (2021): 714186, https://doi.org/10.3389/fnagi.2021.714186.

33. Alan L. Whone et al., "Extended Treatment with Glial Cell Line-Derived Neurotrophic Factor in Parkinson's Disease," *Journal of Parkinson's Disease* 9, no. 2 (2019): 301–313, https://doi.org/10.3233/JPD-191576.

34. Peng Chen et al., "Basic Fibroblast Growth Factor (bFGF) Protects the Blood–Brain Barrier by Binding of FGFR1 and Activating the ERK Signaling Pathway after Intra-Abdominal Hypertension and Traumatic Brain Injury," *Medical Science Monitor: International Medical Journal of Experimental and Clinical Research* 26 (2020): e922009-1, https://doi.org/10.12659/MSM.922009.

35. Justin T. Rogers et al., "Reelin Supplementation Enhances Cognitive Ability, Synaptic Plasticity, and Dendritic Spine Density," *Learning & Memory* 18, no. 9 (2011): 558–564, https://doi.org/10.1101/lm.2153511.

36. Francisco Lopera et al., "Resilience to Autosomal Dominant Alzheimer's Disease in a Reelin-COLBOS Heterozygous Man," *Nature Medicine* 29, no. 5 (2023): 1243–1252, https://doi.org/10.1038/s41591-023-02318-3.

37. Hansruedi Mathys et al., "Single-Cell Atlas Reveals Correlates of High

Cognitive Function, Dementia, and Resilience to Alzheimer's Disease Pathology," *Cell* 186, no. 20 (2023): 4365–4385, https://doi.org/10.1016/j.cell.2023.08.039.

38. Hansruedi Mathys et al., "Single-cell Multiregion Dissection of Alzheimer's Disease," *Nature* 632 (2024): 858–868, https://doi.org/10.1038/s41586-024-07606-7.

39. Miguel Ramalho-Santos and Holger Willenbring, "On the Origin of the Term 'Stem Cell,'" *Cell Stem Cell* 1, no. 1 (2007): 35–38, https://doi.org/10.1016/j.stem.2007.05.013.

40. Samantha Lyons, Shival Salgaonkar, and Gerard T. Flaherty, "International Stem Cell Tourism: A Critical Literature Review and Evidence-Based Recommendations," *International Health* 14, no. 2 (2022): 132–141, https://doi.org/10.1093/inthealth/ihab050.

41. Dominique S. McMahon, "The Global Industry for Unproven Stem Cell Interventions and Stem Cell Tourism," *Tissue Engineering and Regenerative Medicine* 11 (2014): 1–9, https://doi.org/10.1007/s13770-013-1116-7.

42. Xin-Yu Liu, Lin-Po Yang, and Lan Zhao, "Stem Cell Therapy for Alzheimer's Disease," *World Journal of Stem Cells* 12, no. 8 (2020): 787, https://doi.org/10.4252%2Fwjsc.v12.i8.787.

43. Priyanka Mishra et al., "Rescue of Alzheimer's Disease Phenotype in a Mouse Model by Transplantation of Wild-Type Hematopoietic Stem and Progenitor Cells," *Cell Reports* 42, no. 8 (2023), https://doi.org/10.1016/j.celrep.2023.11295.

44. Maurizio A. Leone et al., "Phase I Clinical Trial of Intracerebroventricular Transplantation of Allogeneic Neural Stem Cells in People with Progressive Multiple Sclerosis," *Cell Stem Cell* 30, no. 12 (2023): 1597–1609, https://doi.org/10.1016/j.stem.2023.11.001.

45. Petrou Panayiota et al., "Effects of Repeated Autologous Mesenchymal Stem Cells Transplantation on Cognition and Serum Biomarkers in Progressive Multiple Sclerosis: Interim Analysis of an Open Label Extension Trial," *Multiple Sclerosis Journal* 29 (2023): 1066–1067.

46. Abigail Beaney, "Alzheimer's Trial Doses Patients with Stem Cell Treatment Directly to Brain," Clinical Trials Arena, April 4, 2024, https://www.clinicaltrialsarena.com/news/alzhimers-trial-stem-cells-directly-brain/.

47. Yuancheng Lu et al., "Reprogramming to Recover Youthful Epigenetic Information and Restore Vision," *Nature* 588 (2020): 124–129, https://doi.org/10.1038/s41586-020-2975-4.

48. Jae-Hyun Yang et al., "Loss of Epigenetic Information As a Cause of Mam-

malian Aging," *Cell* 186, no. 2 (2023): 305–326, https://doi.org/10.1016/j.cell.2022.12.027.

49. Jae-Hyun Yang et al., "Chemically Induced Reprogramming to Reverse Cellular Aging," *Aging* (Albany, NY) 15, no. 13 (2023): 5966–5989, https://doi.org/10.18632/aging.204896.

INDEX

Page numbers in *italics* refer to tables.

Ackerley, Mary, 224
Adderall, 10, *304*
adenosine triphosphate (ATP), 32, *291*
ADLs (activities of daily living), 15–16
Aduhelm (aducanumab), 76
Advanced Cognitive Training for Independent and Vital Elderly (ACTIVE), 203
advanced glycation end products (AGEs), 32–33, 35
Allen, Jennie, 211
alpha-synuclein, 244, *284*, 311
Alzheimer, Alois, 97
Alzheimer's disease, 3–6, 11–12, 14–15
   Big Pharma and, 74–81
   early-onset, 19–22, 275
   fasting and, 145
   growth factors and, 272–273
   identification of, 97
   inflammation and, 28
   neuroplasticity and, 25–29, 199–202, 208
   Parkinson's disease and, 243, 244
   peptides and, 270–271
   risk factors for, 27
   stages of dementia and, 14–16
   supplements and, 133–135, *298*, *300*
   testing for, *283*, *284*, *290*
   toxins and, 221, 222, 224, 232–233
   treatability of, 199–200, 309
   See also amyloid beta; amyloid plaques; ApoB; brain-derived neurotrophic factor; p-tau
amyloid beta
   Alzheimer's diagnosis and, 76
   Alzheimer's drugs and, 75, 204
   antiamyloid antibody trials, 204
   inflammation and, 28, 29, 33–34
   insulin production and, 33–34
   stem cell therapies and, 277
   testing for, 175, *283*, *284*, *286*, *290*
   therapeutics and, *296*, *304*
amyloid-beta 42:40 ratio, 114, *283*, *284*
amyloid plaques, 113, 275
amyloid precursor protein (APP), 46, 232–233, 271–272
amyotrophic lateral sclerosis (ALS), 18, 23–25, 43, 97, 115
Ananich, Jim, 229
Anaplasma, 255, *289*
anesthetic agents, 28, 35, 220, 313, 314

antagonistic pleiotropy, 23, 29, 31
antihypertensives, 79, *293*
anxiety, 28, 36, 73, 142, 145, 181, 188, 195
aortic pulse wave velocity, 170
aphasia, 98, 175
ApoB, 85–86, 108, *286*, 308, 311
ApoE 3/4, 307
ApoE4, 50–51, 146, 150, 175
　informational resources for, 221
　macular degeneration and, 310
　target values for, *286*, *292*
　therapeutics and, *304*
　two copies of, 50, 51, 75, 250
apolipoprotein E, 50. *See also* ApoE4
artificial intelligence (AI), 26, 83
Asimov, Isaac, 124–125
Aspergillus, 99, 222, 226, 309
atrial fibrillation, 27, *295*, 314–315
autoantibodies, 45, *287*
Averbakh, Yuri, 61, 65
Ayurvedic medicine, 3, 248

Babesia, 255, 257, *289*
babesiosis, 256, 257
baby boomers, 6
bariatric surgery, 90
Bartonella, 99, 193, 255, 258, *289*
Barzilai, Nir, 273
basic fibroblast growth factor (bFGF), 274–275
Benson, Frank, 98
Benson's syndrome, 98
beta-hydroxybutyrate, 107, 144, 155, 308
bio-identical hormone replacement therapy (BHRT), 264–268, *304*, 309
biotin (B7), *294*
biotoxins, 28, 35, 221–229
　detoxification goal, *293*
　mitigation for exposure to, 124
　mold and mycotoxins, 99, 130, 157, 222, 223, 224, 225–229, *289*, 309
　nicotine, 222
　testing for, *289*
　tetanus, 222
blood flow restriction (BFR), 165–167, 173, 262
blood flow restriction training (BFRT), 165–168, 173
blood sugar levels, 144, 148, 191, 264
blood-sugar monitoring, 80

Blue Cross Blue Shield Association, 19–20
blue light, 25, 182–183
body mass index, 85, 132, 226, *285*, 311
Borrelia, 255, 257, *289*
BrainCheck, 118
brain-derived neurotrophic factor (BDNF), 27, 46, 60, 233, 272, *288*, *303*
brain fog, 11, 37, 39, 169, 251, 252, 265
BrainHQ, 204, 206, *306*
brain training, 221, *293*, *306*, 309, 313
　emotions and, 214–218
　neuroplasticity and, 201–206
　social connection, 210–214
　video games and, 61
Brenes, Gretchen A., 197
B vitamins, 132–133
　biotin (B7), *294*
　choline (B4), 27–28, 124, 132, 137
　cobalamin (B12), 124, 132, 133, 247, *285*
　folate (B9), 124, 132, 137, 247, *285*
　niacin (B3), 247
　pantothenic acid (B5), 247
　pyridoxine (B6), 124, 247, *285*
　riboflavin (B2), 132, 247
　thiamine (B1), 60, 124, 132, 247, *294*

California Institute of Technology, 16–17
Cambridge Cognition, 118
Campisi, Judith, 268–269
Carnahan, Jill, 224
Charcot, Jean-Martin, 97
childhood trauma, 36
cholesterol target values, *286*
choline (B4), 27–28, 124, 132, 137
cobalamin (B12), 124, 132, 133, 247, *285*
Coca-Cola, 126, 143
cocaine, 10, 125
Cognitive Neuroscience Society Vital Signs (CNSVS), 117–118
cognoscopy, 111, 114, 175
Cohen, Deborah A., 157
Connell, Katherine "Kit," 61, 65
continuous positive airway pressure (CPAP), 186, *305*
Cooper, Kenneth, 153
corticosteroids, 91
cortisol, 31, 47–48, 177, 268, *293*
Covey, Stephen, 93
COVID-19, 37–39, 55, 103, 169, 212, 228, 250–253, 308

Index

C-reactive protein, 90, 109, *285*
Cunnane, Stephen, 107
cytokines, 38, 90, 244–245, 252, 272–273

Dance Dance Revolution, 172
Danhauer, Suzanne C., 197
davunetide, 270, 271, *307*
dental amalgams, 232, 247, *301*
Dentalcidin, 248, 249, *301*
depression, 28, 37, 73, 142, 195, 216–217, 270, 311
detoxification, 147, 221, 237–240, *293*, 309, 312–314
    cruciferous vegetables and, 123, 234
    glutathione and, 123, 239, *298*, 312, 313, 314
    saunas and, 239
    sleep and, *293*
    supplements, *298–299*
    too-rapid detox, *299*
diabetes, 4, 33–34, 79–80
    Alzheimer's disease and, 34
    diet and, 126, 138
    prediabetes, 34, 79
    pre-prediabetes, 34, 79
    semaglutides for, 147
    statins and, 86
    testing for, 106
    type 2 diabetes, 27, 34, 79, 106
diet and food, 120–126, *293*
    KetoFLEX 12/3, 121–126, 188
    plant-based food and diets, 85–86, 126–130, 144, 233, 245, 262
    probiotics, 141–142
    processed foods, 137–141
    protein, 130–132
    semaglutides and, 147–148
    SMASH fish (salmon, mackerel, anchovies, sardines, herring), 130–131, 191
    vegan diets, 85–86, 126–127, 130, 132
    vegetarian diets, 85, 124, 126
    See also fasting; ketosis and ketogenic diet; nutrients; sugar; supplements
dihexa, 270
Duke University School of Medicine, 17

early-onset dementia, 19–22, 275
Eastwood, Jessica, 141
Edwards, Marc, 230

Ehrlichia, 255, *289*
Eisai, 75–77, 79, 80
electromagnetic fields, 192–195
emotional plasticity, 214–218
energetics, 27–29, 40–43, 104, 122
    diet and, 124, 262, *293*
    fasting and, 145–146
    glucose metabolism and, 89, 105
    memory and, 41–42
    microbiomes and, 244
    NAD+ and, 263
    oral health and, 248
    oxygen therapy and, 169, 171
    psychiatric diseases and, 97–98
    sleep apnea and, 185
    strength training and, 159
    supplements and, 136, 161
    toxins and, 233, 234
Environmental Relative Moldiness Index (ERMI), 228, *289*
epigenetics, 36, 109, 145, 258, *291*
Epstein-Barr virus (EBV), 45, *289*
evolution
    cortisol and, 47–48
    early health selection and, 46–47, 83
    energy conservation and, 164
    food availability and, 147
    mold and, 223
    MS and, 44
    neuroplasticity and, 25–27
    performance selection and, 22–25, 29, 32, 125
    sleep and, 182
    sugar preference and, 32
    trophic changes and, 46–47
exercise, physical, 43, 150–153, 172–173, *293*
    aerobics, 153–158
    blood flow restriction training, 165–168
    high-intensity interval training (HIIT), 162–165, 170–173
    sleep and, 187–188
    strength training, 158–162
    VO_2 max, 156
exercise with oxygen therapy (EWOT), 168–172

false hopelessness, 14
fasting, 105–107, 123–124, 145–146, 163–164, 188–189, 269
fentanyl, 82

Feynman, Richard, 10, 23
Field, Andy, 212
fisetin, 269, *304*
Flint, Michigan, water crisis, 229–231
folate (B9), 124, 132, 137, 247, *285*
food. *See* diet and food
Franco, Marisa G., 211
Freud, Sigmund, 97
frontotemporal dementia (FTD), 115

Gehrig, Lou, 24
Gen Z, 6, 150
Gersh, Felice, 264, 266
gingipains, 248–249
gingivitis, 91, 247, 248
Glatt, Ryan, 172
Glial CAM, 45
glial cell line–derived neurotrophic factor (GDNF), 274–275
glial fibrillary acidic protein (GFAP), 110–116, 145, 234, 282–283, *284*, 307–311
GLP-1 agonists, 147–148, *300*
glucose metabolism, 105–107, 145, *290*
glymphatic system, 155–156, 179–181, 183, 189, 195–196, 198, 274
glyphosate, 25, 26, 28, 35, *288*, 312. *See also* herbicides and pesticides
Gray, John Purdue, 81
Gregory, Julie, 169–170
growth factors, 272–276
  basic fibroblast growth factor, 274–275
  brain-derived neurotrophic factor, 27, 46, 60, 233, 272, *288*, *303*
  cytokines, 38, 90, 244–245, 252, 272–273
  glial cell line–derived neurotrophic factor, 274–275
  IGF-1, 273–275
  nerve growth factor, 46, 272, *303*
  reelin, 275–276

Hall, Prudence, 264, 266
Health Effects Roster of Type-Specific Formers of Mycotoxins and Inflammagens (HERTSMI-2), 228, *289*
Hemoglobin A1c (HbA1c), 105–107
Henderson, Victor W., 216

herbicides and pesticides, 25–26, 36, 123, 163, 237, 242
  acrolein, 312
  glyphosate, 25, 26, 28, 35, *288*, 312
  paraqua, 26, 312
Herpes viruses, 28, 247, 254, *289*, *302*
Hertz, Noreena, 211
high-sensitivity C-reactive protein (CRP), 109, *285*
Hindin, Howard, 249
Hippocrates, 136
Hodes, Shirley, 55, 56
HOMA-IR, 124, 307–308, 311
homocysteine, 109, 133, 247, *285*
Hood, Leroy "Lee," 16–17, 19, 218
hormesis, 163, 171
Horowitz, Richard, 69
Human Genome Project, 16–17
human growth hormone (somatotropin), 268
Humburg, Craig, 43
Huntington's disease, 18, 97
hyperbaric oxygen therapy (HBOT), 154–155, *298*
hyperimmunity, 250
hypoglycemia, 33, 189

IGF-1, 273–275
imaging and scans, 116–117, *290*
  cone beam CT, *290*
  MRIs, 14, 99, 116, 159, *290*
  PET scans, 15, 33–34, 116, 175, 282, *290*
  volumetrics, 99, 116–117, *290*
inflammation, 89–92, 109–112, 115–116
  amyloid beta and, 28, 29, 33–34
  anti-inflammatory drugs, 90–91
  anti-inflammatory therapeutics, *296*, *302*, *303*
  cognitive impairment and, 90
  Lyme disease and, 91, 256
  neuroplasticity and, 28–29
  obesity and, 90
  sleep apnea and, 124
  supplements and, *301–303*
  target values, *285–287*
insomnia, 28, 162. *See also* sleep
insulin-degrading enzyme (IDE), 33–34
insulin resistance, 33–34, 105–106, 123, 124, *285*, 307–308

Index

Insys Therapeutics, 82
International Society for Environmentally Acquired Illness (ISEAI), 224

Johnson, Richard, 34–35

KAATSU, 165–167, *293*, 308
Kandel, Eric, 215–216
Kelly, Lance, 5
kerygmachela, 182
KetoFLEX 12/3, 121–126, 188
ketones, 107–108, 144–146, *297*
ketosis and ketogenic diet, 95–96, 107–108, 122–124, 144–148, 158, 159, 188, 221, 262, *293*, *297*
Kurkinen, Markku, 75–76

La Crosse encephalitis, 258
Lance, Colleen, 191
Lazar, Sara W., 196–197
leaky gut, 28, 91, 253, *287*, 310
Leqembi (lecanemab), 75–76, 80
Levi-Montalcini, Rita, 46
Lewy body dementia, 243, 283, *284*, 311–312
lifespan. *See* longevity, human
Living Proof (documentary), 70
Long COVID, 39, 169. *See also* COVID-19
longevity, human, 50–59, 65–67
  complex problem-solving and, 60–61
  creativity and, 64–65
  emotional connections and, 62–63
  emotional stability and, 63–64
  high-level reasoning and, 62–63
  learning new skills and, 61–62
  sharp memories and, 59–60
Longo, Valter, 145
Lou Gehrig's disease. *See* amyotrophic lateral sclerosis
L-tryptophan, 131, 190–192
Lustig, Robert, 33, 71
Lyme disease, 25, 91, 255–257, 258–259
  chronic Lyme disease (CLD), 69–70
  inflammation and, 91, 256
  testing for, *289*

Maczulak, Anne, 245
magnetic resonance imaging (MRI), 14, 99, 116, 159, *290*

Markoff, Betty, 62–63, 65, 72
Markoff, Morrie, 62–63, 65, 72
Marshall, John, 183
meditation, 195–197, 308, 309
melatonin, 182, 190–191, 267, *305*
Memories for Life—Reversing Alzheimer's (documentary), 70
mercury, 27, 28, 35, 248, *301*
  air pollution and, 231
  Alzheimer's disease and, 130, 232–233
  cognitive decline and, 35, 230
  in dental amalgams, 232, 247, *301*
  diet and, 122, 130–131, 232–233
  energetics and, 27, 28
  testing for, *288*
Merzenich, Mike, 201–202, 204–205
metabolic syndrome, 28, 34
metformin, *300*
microbiomes, 241–245
  gut health and, 245–247
  oral health and, 247–249
  pathogens and, 257–259
mild cognitive impairment (MCI), 15–16
  GFAP and, 111
  ketone supplementation and, 107
  MoCA scores and, 12, 169, 175, 251
  ReCODE Protocol and, 38, 204, 219
  strength training and, 158
Milman, Sofiya, 273
Mini-Mental State Examination (MMSE), 53–54
mitochondria, 23–24, 26–27, 32, 41, 60, 136, *293*, 312–313
mold and mycotoxins, 99, 130, 157, 222, 223, 224, 225–229, *289*, 309
Montreal Cognitive Assessment (MoCA), 11–12, 14, 15, 53–54, 117–118, 169, 175, 236, 251, *291*
Mosconi, Lisa, 265
motor neuron disease. *See* amyotrophic lateral sclerosis
Mounjaro, 186
multiple sclerosis (MS), 44–45, 70, 97, 115, 145, 222
mutations, 2–3
  aging and, 31
  ALS and, 25
  Alzheimer's disease and, 46
  reelin and, 275
  testing for, *292*

mycotoxins, 99, 130, 157, 222, 223, 224, 225–229, *289*, 309
myoclonus, 174–175

NAD+, 92, 263, *297*
Nakamura, Hikaru, 207
Nasreddine, Ziad, 12
Nathan, Neil, 99, 192–193
National Confectioners Association, 143
Nedergaard, Maiken, 155–156
nerve growth factor (NGF), 46, 272, *303*
neuroplasticity, 25–26, 28, 154, 159, 166, 199–205
   brain stimulation and, 208–209
   brain training programs and, 201–206
   cognitive challenge and, 205–208
   depression and, 217
   emotions and, 214–218
   evolution and, 25–27
   inflammation and, 28–29
   reelin and, 276
   social connection and, 210–214, 262
   stress and, 28–29
   therapeutics and, *294*
   toxins and, 28
NfL, 115–116, 251–252, 283, *284*, 308, 311
niacin (B3), 247
Nicholls, David G., 24
Nietzsche, Friedrich, 163
nonsteroidal anti-inflammatory drugs, 90–91
Norwitz, Nichola, 98
nutrients, 132–136, 137
   CoQ10, 136
   magnesium, 135
   omega-3 fatty acids, 91, 122, 124, 134–135, 232
   target values, *287*
   vitamin C, 133
   vitamin D, 133–134
   zinc, 135–136
   See also B vitamins; supplements

obesity, 21, 89–92, 157, 185
omega-3 fatty acids, 91, 122, 124, 134–135, 232
omega-6 fatty acids, *286*
opioid crisis, 82

oral health and hygiene, 28, 39, 91, 247–249, 259. *See also* dental amalgams
oral secretagogues, 268
oxidative stress, 145, 160, 230
OxyContin, 81, 82
oxygen saturation (SpO2), 41, 103, 155, 176, *291*, *293*, 308
Ozempic, 147

Palmer, Christopher, 97–98
panspermia hypothesis, 22
pantothenic acid (B5), 247
paraqua, 26, 312. *See also* herbicides and pesticides
Parkinson's disease, 18, 23, 25–26, 242–244, 311–314
   Alzheimer's disease and, 243, 244
   energetics and, 146
   fasting and, 145
   GDNF and, 274
   identification of, 97
   metformin and, *300*
   respiratory complex I and, 26
   testing for, *283*, *284*, *290*
pathogens, 37–39, 257–259, *289–290*, *291*. *See also* COVID-19
Patient Zero, 176
PE 22-28, 270
Peper, Erik, 138–139
peptides, 28, 33, 90, 269–271, *307*, 312
Peraica, Maya, 226
Perak, Amanda, 108
periodontal disease, 39, 91
Petersen, Melody, 77
Philip Morris International, 137–138
pinealon, 270, *307*
plant-based food and diets, 85–86, 144, 233, 245, 262
Pollan, Michael, 139
Porphyromonas gingivalis, 248–249
positron emission tomography (PET), 15, 33–34, 116, 175, 282, *290*
Potts, Pauline, 153
Powassan virus, 256
pregnenolone, 91, 268, *287*, *305*
presymptomatic stage of cognitive decline, 15
primary progressive aphasia (PPA), 98

Index

prion disease, 18, 111
progressive supranuclear palsy (PSP), 25, 271
prostate health, 162, 314
protection, target values, *285–287*
p-tau, 15, 113–116, 244, 282–283, 307–311
p-tau 217, 15, 112–116, 145, 217, 251, 282–283, *284, 285,* 307, 308, 311
p-tau 217 test, *284*
Purdue Pharma, 81–82
Purdy, Joe, 177–178
purpose and long-term goals, 93–101
pyridoxine (B6), 124, 247, *285*

Quiet Epidemic, The (documentary), 69–70

ReCODE Protocol, 5, 38, 49, 99, 159, 167, 170, 179–180, 200–201, 206, 212, 219–220, 222, 271
reelin, 275–276
Reno, Hilary, 253
Revitin, 249
riboflavin (B2), 132, 247
R. J. Reynolds Tobacco Company, 138
Rocky Mountain spotted fever, 256
Ross, Jaime M., 234
Ross, Mary Kay, 224
Rubin, Leah, 115
Rutland, Kerry Mills, 98–99
Rybelsus, 147

Sacks, Stanley, 62, 65
Saint Louis University Mental Status (SLUMS), 117, *291*
Satizabal, Claudia, 134
Sato, Yoshiaki, 165–166
Savino, Les, 54–55, 56
Schreiner, Thomas, 184
screens and screen time, 21, 181–182
semaglutides, 147–148, *300*
senescence-associated secretory phenotype (SASP), 268–269
"senior moments," 13
senolytics, 268–269, *304*
Serhan, Charles, 91
Sethi, Shebani, 98
sexually transmitted infections, 253–255
Shanahan, Lilly, 109

shinrin-yoku (forest bathing), 197
sickle cell anemia, 168
Sinclair, David, 258–259
single-photon emission computed tomography (SPECT), 116
sleep, 174–177, *293*
   bedtime and wake time, 183
   deep sleep, 175, 179–180, 183, 187, 189–190, 208, *293*
   electromagnetic fields and, 192–193
   exercise and, 187–188
   food and, 188–190
   glymphatic system and, 179–181, 183, 189, 195–196, 198
   insomnia, 28, 162
   light and, 182–183
   medications and supplements, 190–192
   meditation and, 195–197
   melatonin and, 182, 190–191, 267
   oxygen saturation (SpO2) during, 176, *291*
   REM sleep, 179–181, 186, 189, 190, *291, 292, 293,* 308, 311
   screens and, 181–182
   stress and, 177–178, 188, 193, 195, 197–198
   tryptophan and, 190–192
sleep apnea, 185–187
   atrial fibrillation and, 314–315
   cognitive compromise and, 156, 185–186
   diet and, 124
   energetic supply and, 41, 248
   inflammation and, 124
   oral health and, 248
   risk factor for Alzheimer's disease, 27
   tests for, *291*
   treatments for, 186, 305
sleep study, 186, *291*
Smith, A. David, 133
Snegov, Sergey, 104
social connection, 210–214, 262
socioeconomics, 86
spinal muscular atrophy, 18
spinocerebellar ataxia, 18
Stachybotrys, 99, 222, 309
stages of dementia, 14–16
standard American diet (SAD), 125
statins, 86, 136, *293*

StellaLife, 248, *301*
stem cells, 276–279
stilbenoids, 128
stool analysis, *291*
stress, 122, 308–311, 313
   brain training and, 204
   cortisol and, 31, 47–48, 177, 268, *293*
   fasting and, 163–164
   fight or flight or faint response to, 193
   glympathic system and, 193, 195
   managing and reducing, 63–64, 90, *293*
   neuroplasticity and, 28–29
   physical symptoms of, 103
   probiotics and, 141
   sleep and, 177–178, 188, 193, 195, 197–198
   sound and, 47–48
strokes, 85, 97, 208
subjective cognitive impairment (SCI), 15–16, 219, 282, 309
sugar, 32–36, 142–144
   Alzheimer's disease and, 33–36
   lower glycemic sources of, 129–130
   oral health and, 249
   in orange juice, 121
   performance and, 10
   in standard American diet, 125
   in ultraprocessed foods, 138
   See also glucose metabolism
Sugar Research Foundation, 142
supplements, *294–305*
   BHRT, *304–305*
   detox, *298–299*
   energy and structure, *297–298*
   glycemic control, *300*
   gut health, *299*
   herbs and related, *296–297*
   inflammatory, immune, and antimicrobial, *301–303*
   minerals, *295–296*
   oral care, *301*
   vascular optimization, *300–301*
   vitamins, *294–295*
Sweedler, Ruth, 55, 56
Syn-One, 283, *284*, 311, 314

Tanio, Craig, 224
testosterone, 161–162, 268, *287*, *304*, *305*
thiamine (B1), 60, 124, 132, 247, *294*
thymosin alpha-1, *307*
thymosin beta-4, *307*
thyroxine, 268, *305*
tick-borne pathogens, 255–257. *See also* Lyme disease
Tillis, Thom, 226
tirzepatide, 186, *300*
tobacco companies, 137–139
Toups, Kat, 224
toxins, 220–225
   Alzheimer's disease and, 221, 222, 224, 232–233
   anesthetic agents, 28, 35, 220, 313, 314
   dementia and, 35–37
   metals, 229–233
   microplastics, 233–235
   mycotoxins, 99, 130, 157, 222, 223, 224, 225–229, *289*, 309
   testing for and target values, *288–289*
   volatile organic compounds (VOCs), 221, 225, 235–237, 239
   See also biotoxins; detoxification; herbicides and pesticides; mercury
traumatic brain injury (TBI), 155
Treponema denticola, 248
trichloroethylene (TCE), 26
triiodothyronine, 268
trophic factors, 27, 28, 34, 46–47
trophic support, *287–288*
tryptophan, 131, 190–192
Tucker, Howard, 55, 56

ultraprocessed food, 71, 86, 125, 138, 140
ultraviolet light, 33, 128

Val66Met, 233
Valenzuela, Michael J., 158
vascular health, target values for, *285–287*
vegan diets, 85–86, 126–127, 130, 132
vegetarian diets, 85, 124, 126
video games, 61, 172
volatile organic compounds (VOCs), 221, 225, 235–237, 239
volumetrics, 99, 116–117, *290*

Index

waist to hip ratio, *285*
Walter, Catherine, 60, 65
Warren, Elizabeth, 226
washing machines, 194–195
wearable health monitors, 102–103, *291*, 308, 311
Wegovy, 147
West Nile virus, 258
Williams, George C., 23, 29
Winfrey, Oprah, 94

World War I, 60
World War II, 56, 58, 61, 72, 153

Yang, Jae-Hyun, 279
Ybarra, Oscar, 213
yoga, 196–197

Zika virus, 258
zombie cells, 268–269

ABOUT THE AUTHOR

Dr Dale E. Bredesen is internationally recognized as an expert in the mechanisms of neurodegenerative diseases such as Alzheimer's. He graduated from Caltech, then earned his MD from Duke University Medical Center. He served as chief resident in neurology at the University of California, San Francisco, before joining Nobel laureate Stanley Prusiner's laboratory at UCSF as an NIH postdoctoral fellow. He has held faculty positions at UCSF, UCLA, and UCSD. Dr. Bredesen directed the Development, Aging and Regeneration Program at Sanford Burnham Prebys before moving to the Buck Institute in 1998 as its founding president and CEO. He is currently senior director of the Precision Brain Health program—the first program of its kind—at the Pacific Neuroscience Institute in Los Angeles.